ENGINEERING IN THE
ANCIENT WORLD

ENGINEERING IN THE ANCIENT WORLD

J. G. Landels

University of Reading

University of California Press
Berkeley and Los Angeles

University of California Press
Berkeley and Los Angeles, California

This revised edition first published 2000

First published 1978. First paperback printing 1981

Copyright © J. G. Landels 1978; 2000

Library of Congress Catalog Card Number: 76-52030

ISBN 0-520-22782-4

09 08 07 06 05 04 03 02 01 00

10 9 8 7 6 5 4 3 2 1

Printed and bound in the EU

Contents

Preface to the second edition

THE purpose of this book is to discuss and illustrate a number of technological achievements in the Greek and Roman world. Twenty years ago most of the information on these topics was contained (if anywhere at all) in a number of highly specialized studies, not all of them easily accessible, and few of them written by classical scholars. What I attempted to do was 'to give the reader (whether a student of classical civilization or a layman interested in the history of engineering) some insight into the mechanical skills of the two most fascinating civilizations of ancient Europe'.

In the twenty years since the publication of the first edition there has been a considerable upsurge of interest in ancient technology, and a number of major works have been published, including an important source-book. It has been fully recognized that arguments based on technical constraints which remain today exactly as they were two thousand years ago can be useful in solving some historical problems—an approach which was pioneered by the late Professor J.E. Gordon. There have, accordingly, been a number of projects carried out, some of them involving a fruitful collaboration between classicists, technologists and archaeologists (in whichever order of priority the reader considers appropriate). It is no longer justifiable therefore to complain, as I did in the preface to the first edition, that 'in most standard histories, the archaeological evidence is treated in a descriptive way, and very little attempt is made to envisage mechanical contrivances in action', or that 'the written sources are not always examined in detail, and the Greek and Latin terminology is not usually analysed.'

The main text of the first edition has been reprinted without change. The translations of passages quoted from Latin and Greek authors were all my own; other versions may be superior in literary merit, but even the most eminent translators can sometimes make extraordinary mistakes on technical matters, such as the techniques of boat-building used by Odysseus in *Odyssey* V. Two sections have been added at the end—an appendix (pp. 219–24) on the reconstruction of the trireme which was completed in 1987, and a section containing some further thoughts, mostly on matters to which my attention was drawn by reviewers and correspondents. Also the bibliography has been expanded and updated, so far as time and space have allowed.

In writing the first edition I was helped by a number of scholars in various disciplines, and here I must acknowledge once more my indebtedness to Professor J.E. Gordon, who died in 1998. He allowed me to see his notes on a course of lectures on the history of naval architecture (the graph on p. 167 was reproduced with his kind permission), and gave much helpful advice on the engineering problems of catapults. I also received much useful information on archaeological finds from Dr (now Professor) Michael Fulford.

More recently, Dr Michael Lewis has given valuable help with bibliographical references, and Mr Digby Stevenson has kept me informed of his researches on catapult spring materials. But my most important debt is to Dr Boris Rankov, who has been most helpful in giving prompt and expert answers to my questions on rowing matters, and has kindly allowed me to see a draft of his chapter on the sea trials of *Olympias* which is due to appear early in 2000 in a new edition of Morrison and Coates' *The Athenian Trireme* (CUP).

Twenty years later, I have once again to thank my wife Jocelyn for her help and encouragement.

Reading, September 1999   J.G. Landels

1

Power and energy sources

THE sources of power available in classical antiquity were severely
limited by comparison with those of the present day. Virtually all
work was done by man-power or animal power, and the kind of
constraint which this imposed may be seen from a simple illustration. One gallon of petrol may seem very expensive nowadays, but
if used in an ordinary engine of average efficiency it will do the
equivalent work of about 90 men, or of nine horses of the smallish
size used in the ancient world, for one hour. Water power was
used for pumping and industrial purposes, but probably not much
before the first century B.C. The theoretical possibilities of steam
power, hot air expansion and windmills were known, but apparently never exploited except on a very small scale, and not in useful or practical applications.

MAN-POWER

The most common mode of employing man-power was in the
handling and porterage of small burdens of the order of 20–80lb
(9–36kg). This is discussed in detail in Chapter 7, and all that
needs to be done here is to note a very important limitation, which
should be quite obvious, but is all too often forgotten. If a burden
requires more than one man to handle it, its size and shape must
be such as to allow the necessary number of men to stand close
enough and get a grip on it. For example, in the fifth century B.C.
the columns of Greek temples were built up from a number of
sections, called 'column drums'; these might be anything up to 6ft
6in (2m) in diameter. The only possible place to grip such a lump
of stone is around the lower edge, and it would be very difficult for
more than 18 men to get into position to grip it at once. It follows,
therefore, that if its total weight was more than about a ton (as it
often was), they would be unable to lift it up off the ground, let
alone move it, turn it round or position it on a column. They
might just be able to roll it along level ground on its edge, but that

would be all. When, therefore, people say 'of course they had thousands of slaves to do the building for them', two facts should be remembered. Though the Pharaohs in Egypt may have had vast resources of manpower, Greek and Roman building contractors rarely had more than a small labour force, and in any case, no matter how many they may have assembled for the more ambitious projects, they could never have man-handled the larger stones used in classical buildings. Either one of the lifting devices described in Chapter 4 must have been employed or else the very slow and extravagant method of building a ramp, and dragging the stones up the slope on rollers.

There were two important mechanical devices for harnessing man-power. One was the capstan or windlass, particularly useful on cranes or aboard ship. The power could be transmitted over a distance by ropes, its direction could be changed by pulleys, and the force could be multiplied by block-and-tackle arrangements. The windlass itself has a built-in mechanical advantage. It was also found to be ideal where traction was required, of low power but finely and accurately controlled. Two medical uses illustrate this. One was the so-called 'bench of Hippocrates'—a plinth with a windlass at each end to provide the extension needed for reducing fractures and dislocations of the arms and legs. The other was a device apparently used by some gynaecologists—a small capstan mounted below a 'midwifery stool', used for extracting a foetus from the uterus.*

It is generally agreed that the Greeks and Romans did not, apparently, discover or use the crank in place of the handspikes on a windlass. Hero of Alexandria speaks of something called a 'handholder' (cheirolabe) for turning axles. This might have been a crank, but there is no proof that it was. There was at least one situation in which the main advantage of the crank could have been exploited, and where its disadvantage would not have been noticed—the repeater catapult. Since it was not used on that weapon, it seems almost certain that it was not known to the designers.

How serious a drawback was the lack of this device? The answer seems to be—rather less than is sometimes suggested. The only real advantage of using a crank is speed. A single grip (firm, but loose enough to allow the handle to turn in the palms of the hands)

*Hippocrates, *On Joints* chapter 72: Soranus, *Gynaeceia* XXI, 68.

can be maintained all the time, whereas with handspikes the grip has to be changed, usually four times per revolution. But in situations where speed is less important, the crank has a positive disadvantage. The force which can be applied to it varies according to its position in relation to the operator, reaching a minimum twice during each revolution when the handle crosses a line drawn through the operator's shoulders and the axis of the crank, and a maximum when it is roughly at right-angles to that line. This is why a car starting-handle used to be be so arranged that the points at which most force is needed to turn it occurred when the handle was at 'two o'clock' and 'eight o'clock'. This imposes a serious limitation on the crank. When it is under continuous loading (e.g. on a crane when the load is raised, or a well-head when the bucket is full), the reverse thrust applied to the crank handle by the load must never exceed the minimum applied by the operator at the two weakest points of the cycle. If it does, the handle will fly backwards, and once it has started swinging round the load may acquire momentum and make the handle impossible to stop. To avoid this danger, most modern cranked winches are fitted with a ratchet. Such a device, dating from the late fifth century B.C., was found near some naval installations at Sunium, and may have been used on a winch for hauling ships up slipways.

The implications for ancient devices worked by handspikes are clear enough. For cranes or hoists of any kind the use of a crank would have lowered the handling capacity by some 20–30%, and it seems rather improbable that a slight increase of speed would justify that sacrifice. On the repeater catapult the slider was fitted with pawls and a ratchet, and would only fly forward a short distance if the tension on the draw-back cord were relaxed. It would therefore have been reasonably safe to use a crank on the capstan at the rear of the machine and thereby speed up the loading operation—a particularly important benefit, for that particular weapon.

The other mechanical device was the treadmill—a pair of vertical wheels with treads (like those of a step-ladder) between them. It has become very difficult nowadays to talk, or even to think about this apparatus unemotionally, and in purely engineering terms, but in fact, if well designed, it can be one of the most efficient devices for this purpose, and the most comfortable for the operator—in so far as any continuous, monotonous physical work can be comfortable. The basic action is not unlike that of pedalling a bicycle, and it is significant that recent attempts to reach the absolute limits of the human body's capabilities, in the development of man-powered flight, have mostly used that arrangement. The difference is that a cyclist pulls on the handlebars, and uses the abdominal muscles as well as the leg muscles; the treadmill operator uses the reaction from lifting his body weight, mainly with the leg muscles.

A very useful feature of the treadmill, especially when used on a crane, is that the torque, which determines the pull on the hoisting-cable, can be easily and accurately adjusted by the operator shifting his position on the wheel. The maximum torque is obtained when the operator treads the wheel at a point on a level with the axle (this can only be done from the outside). If he treads above that point (outside) or below it (inside) the torque is less, and if he stands directly above or below the axle it is zero. Thus the amount of torque required between the maximum and zero, can be obtained by moving forwards or backwards.

This may possibly afford an explanation of a rather mysterious length of wood with notches along one side, found near the Roman water pumps in the Rio Tinto mines. When these pumps (which themselves acted as treadmills) were being used in a series, it would be very important to keep the output of each of them constant, and consistently the same as that of the pumps above and below —otherwise the sumps would either empty or overflow. If this piece of wood was one of two beams supporting a movable handrail, the necessary adjustments for men of different weights working the same pumps at different times could be made by shifting the rail along one or two notches, forwards to reduce output or backwards to increase it.

A second valuable feature of the man-powered treadmill is its mobility. The crane shown on the monument of the Haterii (p. 84) could presumably have been dismantled, and its jib laid horizontally on one or more carts, while the treadmill itself could have been rolled along any reasonably level road (that was also one method used for transporting column-drums). There was, in fact, no other suitable power source available. Wind power is hopelessly unreliable, and a builder would be extremely lucky to have water power available on the site at all, let alone near enough to any particular building. A glance at the later history of cranes shows that the treadmill continued to serve this need right through the Middle Ages and Renaissance, and that the first alternative to be made effectively mobile was steam, as used on railway breakdown cranes. Indeed, the problem is still with us. Owing to difficulties of gearing and transmission the internal combustion engine is not very suitable for large cranes, and the cost of laying supply cables makes it uneconomical to use electricity for anything less than a large and lengthy building project.

The Greeks and Romans also used manpower for the propulsion of virtually all fighting ships. Merchant ships, except for quite small ones, were normally under sail. Warships used sails on long voyages, or while cruising on patrol, but in battle conditions, or during a battle alert, they usually left mast, yard and mainsail ashore, to cut down weight to the absolute minimum, and relied entirely on rowers.

ANIMAL POWER

From remote antiquity there has been a contrast between the working animals used in the Mediterranean area and those used in northern Europe. The predominance of the horse in northern Europe, closely related to climatic and ecological factors, could never have occurred in classical Greece, and did not affect Roman practice to any great extent except in so far as Roman armies came into contact with the peoples of France, Germany and central Europe. The situation in classical Greece is summed up both accurately and poetically by Aeschylus in a passage of his *Prometheus Bound*. The hero, describing his services to mankind, says (lines 462–6)

> 'And I was the first to link oxen beneath the yoke
> With yoke-straps, to be man's slaves, and with their bodies' strength
> Give him relief from the heaviest of his toil;
> And to the chariot-pole I brought
> Horses that love the guiding reins,
> Delight and pride of massive wealth and luxury'.

The slowness and ugliness of oxen (a generic word, meaning 'great knobbly beasties' is used in the Greek original) is contrasted with the speed and elegance of horses. The assertion in the last line, that horses were expensive to buy and maintain, is borne out by the fact that several words denoting social and economic status in classical Greece were connected with horses. The word *hippeus,* referring to a particular income-group, originally meant a man wealthy enough to own his own horse and (in wartime) to fight in the cavalry of the citizen army. In Athens the next lower property-classification was *zeugites,* meaning a man who owned a pair of oxen. The historian Herodotus, wishing to stress the great wealth of a particular family, calls them *tethrippotrophon*—able to maintain a four-horse racing chariot (for entry at the races during the great games at Olympia, Delphi and elsewhere). The 'conspicuous consumption' of such a family must have made a deep impression.

By contrast, a pair of oxen could be fed much more cheaply, on inferior fodder of a kind available in areas of Greece and Italy where the pasture was not adequate to support horses. They yielded a return on the owner's investment; they could pull a heavier load than two horses of comparable size. Their progress was slower, but then speed was not the most important consideration in ancient farming or transport. Farm animals had to be fed all the time, whether in use or not; a transport contractor would naturally want to complete each job as soon as possible to be ready for the next. But to use horses to speed up his operations would have been quite impractical. And finally—an important point for people living close to subsistence level—when their working life was over, oxen could serve as food. The meat would be tough as old boot, no doubt, and would need a long spell in the stewpot, but it would be better than nothing. The Greeks and Romans, for reasons not clearly defined but presumably religious, did not as a rule eat horsemeat.

The one advantage that the horse had over the ox was speed, and it was precisely in those situations where speed outweighed everything else that the horse was used—in warfare and in chariot-racing. The high mobility of the cavalry gave that arm its particular role in battle tactics, and on the race-course a chariot, made as light as possible, and drawn by a matched team of two or four horses, represented the ultimate in speed to the Greeks from the eighth century B.C. onwards, and to the Romans after them.

Oxen, then, propelled the heavy lorries of the ancient world, and highly-bred horses its Aston Martins and its Lamborghinis. Between these extremes of utility and luxury came the small travelling vehicle for passengers or light merchandise, drawn by donkeys or mules. These animals could move rather faster than oxen, but not as fast as horses. They cost a little more to feed (in proportion to their weight and pulling capacity) than oxen, but only about 60–70% of the cost of horses.

The use of animals in transport, and the problems connected with harness, are discussed in Chapter 7. Apart from transport, the use of animal power was rare. In mining operations it seems to have been almost negligible, for obvious practical reasons. Unless there was access via horizontal tunnels ('adits'), it would be very difficult indeed to get animals into or out of a mine, and ancient workings did not normally include entrance edits or any galleries or spaces in which animals could be kept, fed and housed underground. The haulage of ores and spoil seems to have been done exclusively by man-power, using buckets on ropes, and it was extracted via the nearest shaft, not taken along any great distance underground.

Until about the first century B.C. animals were not used in milling. The only type of mill which can be operated by a horse or donkey is a rotary mill, and that invention did hot come into the classical Greek world at all. The so-called Pompeian mill, with a fixed lower stone of conical shape, and a rotating upper stone shaped like an hour-glass was quite certainly designed to be turned by animal power, despite the fact that the space available in some of the buildings for the animals to walk round seems very limited indeed. The earlier 'pushing' type of mill, in which a grinding stone is pushed back and forth over a trough, must have depended on human effort. Such work was sometimes imposed on slaves as a punishment, but at all times it had to be done by someone, and as a punishment it was probably not much more severe than the 'spud-bashing' to which army offenders used to be subjected—a tedious, irksome job which nobody would do from choice. Some illustrations of rotary mills being turned by horses give a highly idealized picture of noble steeds striding around; in real life, the oldest and most broken-down horses and donkeys were put to this kind of work—the last stage on the road to the knacker's yard.

Finally, there is a bizarre invention described in a Latin work written in the latter half of the fourth century A.D., but almost certainly never constructed. The author's name is not known, and the work is usually referred to as *Anonymus De Rebus Bellicis*. Oxen are used to propel a ship (Fig. 1). They walk around in pairs, at opposite ends of a capstan-pole on a vertical axle. Through a gearing system (not described, but clearly a crown wheel and pinion, as used in water-mills) this axle drives a horizontal one athwart the ship, with a paddle wheel on each end—the description of the paddles has some verbal resemblances to Vitruvius' description of an undershot water wheel (X, 5, 1.) We are not told whether the paddle-wheel shaft was higher or lower than the platform on which the oxen walked, but since they were 'in the

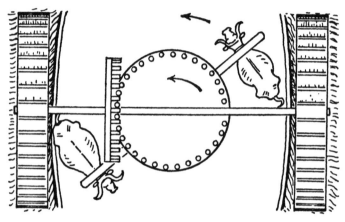

Fig. 1. Oxen used to propel a ship.

hold of the ship', it seems more likely to have been the former. The total number of oxen is not specified, except that there was more than one pair. Though there is no theoretical reason why this should not work, the whole idea does not sound very practical. The space needed for the oxen to move around would be considerable—a circle of 10ft (3m) diameter at the very least. If we assume three capstans, the ship would require a beam of about 13ft (4m) and a length overall of at least 43ft (13m), and at 'six oxpower' such a vessel would be rather under-engined. Communication between the 'bridge' and the 'engine-room' might also be a trifle difficult.

WATER POWER

Early Greek poetry contains striking passages in which the destructive force of rushing water is used as a piece of telling imagery, but the problems of harnessing such power and using it to drive machin-

ery were apparently not explored until the early part of the first
century B.C. According to the geographer Strabo (XII, 3, 40) a
water-mill was built in the kingdom of Mithridates, at Kabeira in
the Pontus (near the modern Niksar, N. central Turkey) in the
first century B.C., some time before the earliest in Greece or Italy.
There may be a simple explanation for this. The basic requirement for a water-wheel is a water supply which is steady all the year
round, and, if it is to be anything more than a toy, the quantity of
water needed is quite large. Mithridates' city was close to a substantial river, the Lycus (modern Kelkit) which, though the local
rainfall is no greater than that of Greece or Italy, has a large catchment area. Relatively few of the rivers and streams of Greece and
Italy (except in the north) maintain a substantial rate of flow during the dry season. However, the effect of this geographical fact
on the history of the water-wheel should not be exaggerated. Once
the basic idea has been put into practice, the conservation and
management of limited or fluctuating water supplies follows soon
afterwards.

Our knowledge of Greek and Roman attempts to harness water
power rests on rather meagre evidence. Among the literary
sources, Vitruvius (late first century B.C.) is much the most important, and he gives a clear description of an undershot wheel,
which is discussed below. Two other allusions are important for
the question of dating. A Greek epigram in the Palatine Anthol-
ogy (IX, 418) speaks of the joyful release from drudgery which a
water-mill has brought to the women servants who previously
had to grind by hand. Its author was almost certainly Antipater
of Thessalonika, who was closely associated with a Roman noble
family, the Pisones. He lived and worked in Italy at the end of
the first century B.C., and is probably referring to the installation
of such a mill on a country estate. His poem would be contemporary with Vitruvius' work, but there is one interesting difference between the two. Antipater speaks of the Nymphs (which
personify the water) as 'leaping down onto the topmost part of
the wheel'. Though this has been disputed, there is really little
doubt that he is talking about an overshot wheel—a more efficient type than Vitruvius' —and this raises a question of priority,
which will be discussed later.

Closely related to this is an allusion in Lucretius' poem *On the
Nature of the Universe*, where the poet is speaking about the move-

ment of the heavenly bodies (V, 509–33, particularly 515–6). It is a difficult and obscure passage, but the gist is that one explanation of the apparent diurnal rotation of the heavens is that a current of air circulates around the universe, causing the 'sphere' to rotate 'as we see rivers turning wheels and buckets *(rotas atque haustra)'*. Since Lucretius uses this as an illustration, he clearly assumes that water-wheels are familiar to his readers, and as he was writing some 40 years earlier than Vitruvius and Antipater, this suggests that the use of water power to work pumps (bucket-wheels or bucket-chains, see Chapter 3) came earlier than its use for milling.

Other literary allusions add little or nothing to this. The archaeological evidence is equally scarce, but very informative. Two important wheel sites have been excavated—one in the Agora at Athens, to the south of where the restored Stoa of Attalus now stands, dating from mid or late fifth century A.D. The other is at Barbegal, near Arles in southern France (just north of the Camargue). A very big installation was built there by the Romans in the late third or early fourth century, and was probably in use for the greater part of 100 years. It contained eight pairs of wheels, each driving millstones in a mill-chamber beside the wheel-pit, and its output would have been adequate not only for the 10,000 inhabitants of Arles, but for some area around. The presence of a Roman garrison might account for this. The remains are not very extensive, but the main essentials can be reconstructed from them. Evidence of an undershot wheel (in the form of chalk incrustation, the wood having all disappeared) has been found at Venafrum in central Italy, and a speculative reconstruction can be seen in the technology section of the Naples Museum.

There are three basic types of water-wheel—the vertical-shaft, the undershot and the overshot. The vertical-shaft wheel has a number of blades inclined at an angle of about 30° to the vertical, fixed to a hub near the bottom of the shaft. The water is directed onto the blades through a wooden trough which slopes down at a steep angle, so that the water strikes them at high speed. This requires a situation where there is a drop of some 10–12 ft (3m) immediately beside the water source. Sometimes a pit can be dug for this purpose, but adequate arrangements have to be made for the spent water to drain away from it. Since the shaft is vertical, it can be made to turn millstones directly, without any need for gears.

In the absence of any conclusive evidence, some historians of the subject have used the following argument: This is the 'most primitive' form of water-wheel, so, since the Romans developed the more sophisticated undershot and overshot wheels (for which we have good evidence), we must assume that they started with the vertical-shaft type. The parallel between this supposed sequence and that attested for Renaissance Europe is also invoked in support. This *a priori* argument is attractive, but it does rest on two doubtful assumptions (a) that milling was the first operation for which water power was used—and the passage from Lucretius quoted above makes this very doubtful—and (b) that gearing of some sort had not been previously invented for other purposes, such as coupling animals to a water-pump. Archaeological evidence (or rather, the lack of it) does not help to decide the question. No certainly identifiable Greek or Roman remains of this type of wheel have been found, but the entire structure, including all the water-guidance system, would have been made of perishable material. By contrast, the overshot wheel required a stone-built wheel-pit, which has good chances of survival, and can be identified as such.

The second basic type of wheel is the undershot, sometimes called 'Vitruvian' from that author's description (X, 5). It is highly significant, and consistent with the evidence from Lucretius, that he first introduces the water-wheel as a power source for working a bucket-chain, and then says, 'It is also used for corn milling, the design being the same except that there is a gear-wheel on one end of the axle . . .' He makes no mention of the vertical-shaft wheel. The structure he describes is very simple (Fig. 2). It consists of a spoked wheel of unspecified diameter, with vanes or paddles around its circumference (Vitruvius calls them *pinnae*, a word used elsewhere to mean the wing-feathers of a bird), which are driven round by the current in the river. There is nothing in his words to suggest that a mill-leat was channelled off for the purpose.

The third and most efficient type of wheel is the overshot (Fig. 3). Using the same kind of argument as with the vertical-shaft wheel, it is usually held that this was developed from the undershot wheel, the intermediate stage in this process being the so-called 'breast-shot' wheel, which is a simple modification of the undershot, the water being supplied through a trough level with the

Fig. 2. Undershot water-wheel.

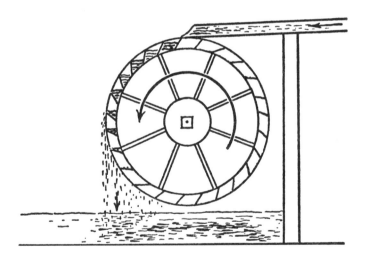

Fig. 3. Overshot water-wheel.

axle, so that the main force on the paddles is from the water falling, not merely flowing past. But there is another equally attractive hypothesis—that the overshot wheel was conceived independently of any other type, by simply reversing the action of the bucket-wheel. If one can put power into that machine and get water out of the top, why not put water into the top and get power out of it? In fact, this possibility would be clearly demonstrated each time somebody finished a spell of work treading a bucket-wheel. It would have to be slowly reversed until all its buckets were emptied, and the pull it exerted during that operation would be clearly felt. The bucket-wheel was certainly in use by Vitruvius' time, perhaps for some while before, and though he does not describe an overshot wheel, that might be due to the fact that the undershot type was the only one he had seen.

The question of priority, then, is not easy to answer; but in power output and efficiency the overshot wheel is well ahead. The structure required for an undershot wheel is simply a vertical wall beside a river or stream,* and, if the water supply is limited, some sort of partial dam to narrow the channel and make it flow more rapidly in the region of the wheel. The potential power output depends on two factors—the velocity of the water flow and the area of the vanes on which the water impinges (the 'scanned area').

To take a simple example. If the area of each vane is 1,000cm² (just over 1 sq ft) we may assume that roughly this area is being scanned at any one time. (The exact figure depends on the number of vanes, the diameter of the wheel, and other factors, but this will do as a crude approximation.) If the water flows past at about 150 cm/see (5ft/sec) the theoretical power available is about $\frac{1}{4}$ h.p. (186 watts), but as the undershot wheel can only be made about 22% efficient at the best, this would provide a real power of only about $\frac{1}{20}$ h.p., or half the power output of a man working a treadmill. If the water flows twice as fast, the power is increased eight times and things look better. The theoretical power available is nearly 2 h.p., and the actual output might be about 0.4 h.p.—the equivalent of four men. On the other hand, the water supply for such a performance could not be obtained from anything less than a small river, with a flow of (say) 125 gall/sec, which might be around 12ft wide (3.5m) and 4in (10cm) average depth in cross-section.

The overshot wheel can be made much more efficient—up to 65% or even 70%. Provided that the wheel revolves fast enough, and the boxes are large enough to catch all the water as it comes from the launder, most of the potential energy in the water can be

*For an example, see the Byzantine mosaic in the Palace of the Emperors, illustrated in *Antiquity* XIII (1939) pp. 354–6 and Plate VII.

utilized. This potential energy can be worked out quite simply from the rate of the water flow, and the depth of fall which, as a rough approximation, may be taken as equal to the diameter of the wheel. The lesser rate of flow given in the last paragraph (31 gall or $140l$/sec), if delivered to an overshot wheel of 7ft (2.13 m) in diameter, would give a theoretical power output of just under 4 h.p. (nearly 3000 watts), and an actual output of perhaps $2-2\frac{1}{2}$ h.p. The power of each of the sixteen wheels at Barbegal might have been of this order. For a modern (and slightly depressing) comparison, a very small motor-cycle engine develops about the same power.

Overshot and undershot wheels may usefully be compared in two other respects—behaviour under extra load and cost of construction. They behave in opposite ways under extra load. Since an undershot wheel is absorbing kinetic energy from the moving water, its torque depends on the difference between the velocity of the water on arrival and the speed of the paddles. To put it very simply, it generates power by slowing the water down. Therefore, if extra loading is put on the wheel (e.g. by using bigger millstones or putting bigger buckets on a chain) it will turn more slowly, but will develop more torque. Conversely, the overshot wheel has a minimum working speed, below which the water begins to overflow the boxes and spill into the pit, reducing the power output and efficiency. These factors must be taken into account when designing the wheel and the gearing, which will be discussed later.

In point of cost, the undershot wheel has the great advantage that no pit is needed, that any riverside situation can be used (this may reduce transport costs, which were high) and no engineering is required to raise the water to the necessary height. This has to be done for an overshot wheel, and may be very expensive. If the gradient of the river bed is slight, it may be necessary to build an artificial channel 200–300 yards long, support it above ground level and make it waterproof. The cost of constructing the aqueduct to feed the Barbegal system must have been very considerable. The efficiency of the undershot wheel is much less, but this need not have worried the ancient engineers all that much. Where fuel is expensive (as in the modern petrol engine) efficiency is the first essential, and must be achieved at almost any cost, but where the energy source is running water, and 'costs nothing', the only requirement of a wheel is that it should deliver enough power to do the work. If the choice lay between an undershot wheel which would just about turn the millstones and an overshot one which would turn them faster, but would cost four or five times as much, and might have to be built some miles away, the undershot would be preferred. Where the water supply was too small for anything less efficient than the overshot, there was no choice. This was probably the case in the Athenian Agora

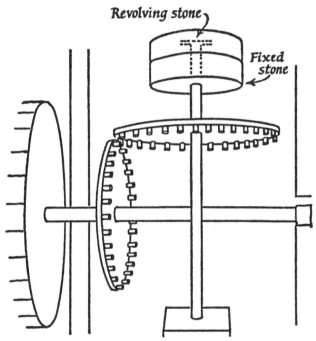

Fig. 4. Water-mill gears with toothed wheels.

It is not difficult to see, from Vitruvius' clear description and from the evidence of the Agora mill, how the water-wheel, turning on a horizontal axle, was coupled to the upper millstone on a vertical axle (Fig. 4.) 'A toothed disc *(dentatum)* is keyed on to the end of the axle, and turns in a vertical plane, at the same speed as the wheel.' (The odd phrase *in cultrum,* which has not been satisfactorily explained, is omitted from this translation.) 'Close to this disc is another larger one, toothed in the same way *(item dentatum)* and horizontally placed, with which it engages *(continetur).'* Vitruvius is clearly talking about two crown wheels. We use the word 'toothed' more loosely, of a flat cog-wheel with radial teeth, but if one thinks of an animal skull, with curving jawbone and teeth at right angles to the 'rim', it will be seen that Vitruvius' usage is really more exact. Marks made by the rim of the vertical gear-wheel on the edge of the gear-pit in the Athenian Agora mill confirm that the teeth were not radial. Vitruvius' expression 'toothed in the same way' *(item dentatum)* suggests that the so-called lantern pinion (Fig. 5) with two discs was not

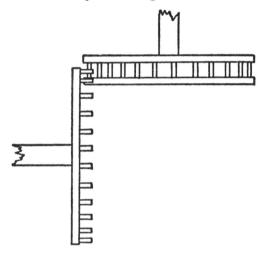

Fig. 5. Water mill gears with toothed wheel and lantern pinion.

known to him; a Roman wood-and-metal pinion of this type has been found in Germany* but what part it played in mill machinery (if any) has not been satisfactorily explained. Doubt has been cast on Vitruvius' statement that the horizontal gear-wheel (coupled to the millstones) was larger than the vertical one on the wheel-shaft, since this would mean that the millstone was geared down, and turned more slowly than the water-wheel. Some scholars have arbitrarily changed the text (from *maius* to *minus)* to avoid this problem. It is true that later European mills had the opposite arrangement, the millstones being geared up by as much as $2\frac{1}{2}$:1, but these were big, powerful overshot wheels, and the type which

*Illustrated in L. A. Moritz, *Grain-mills and Flour in Classical Antiquity* (OUP 1958) Plate 14(c).

Vitruvius describes might not have developed enough power even for a 1:1 gear ratio. The consequences are not nearly so disastrous as some historians suggest. The miller simply worked more slowly, and estimates of flour production should take this into account.

Water-wheels were clearly used for water-raising and for milling. We might expect some other applications, but the only evidence we have is a brief and tantalizing allusion in Ausonius' poem on the River Moselle, written about the middle of the fourth century A.D. Speaking of the River *Erubius* (the Ruwar) he says:

> '*He, turning the millstones with rapid, whirling motion,*
> *And drawing the screeching saws through smooth white stone,*
> *Listens to an endless uproar from each of his banks*' (362–4)

Ausonius' style is not exactly straightforward, and it is difficult to be sure exactly what he means, but he certainly seems to be saying that water-wheels were used to drive saws for cutting stone. The noise was incessant because the power-driven saws, unlike those in an ordinary mason's yard, did not stop for a breather every few minutes. Pliny (*Nat. Hist.* 36, 159) mentions stone from this area and from others which 'can be cut with a saw of the kind they use for cutting wood—even more easily than wood, so they say'. It was used for roof and gutter tiles, and was almost certainly some form of soapstone.

But how did the river 'draw' the saws through stone? *Trahere* would be a strange word to use (even for Ausonius) of a circular saw, though that was perhaps known in antiquity. Did the wheel have a cam and lever, or a crank and connecting-rod to push the saw back and forth? In the absence of any evidence for either we can only guess, and regret all the more that no technical writings have survived from that area or from that period.

Mention was made at the beginning of this section of the destructive power of a river in spate. Though this power itself was not put to useful purpose, some Roman mining installations in Spain, by 'imitating nature', achieved a great saving of manpower and time. Large reservoirs, known as 'hushing tanks', were constructed on the hillsides above the workings, with sluices at one end which could be rapidly opened. When the tanks were filled (in some cases via a fairly long aqueduct) the sluices were released, and a great wave of water rushed over the workings, carrying away with it large quantities of spoil. The same water supply, regulated down to a steady trickle, could also be used for washing ores.

WIND POWER

Although the Greeks and Romans harnessed and used wind power for sailing ships, they do not appear to have developed the rotary windmill as a power source. This is strange, and no satisfactory reason has yet been offered. They were perfectly well aware that by adjusting the set of the sail a boat could be made to travel at an angle to the direction of the wind, and a very slight development of this idea could have led to the type of sail-mill to be seen nowadays on Mykonos and in Crete. But we have no evidence for any such machine in classical antiquity. The one and only mention of harnessing windpower is in the *Pneumatica* of Hero of Alexandria (I, 43), and, being unparalleled, it has come under suspicion as a later interpolation. But there is nothing in the vocabulary or style of the Greek which is inconsistent with the rest of Hero's works, nor is it easy to see what motive could have prompted anyone to insert such a passage into a fairly well-known text some time in the Middle Ages, when the windmill had come into general use.

Hero's machine, in which wind power is used to blow an organ, is crude but workable. Only a very sketchy outline of the instrument itself is given, with no mention of a keyboard or air reservoir. This may mean that it was something like an 'Aeolian harp' (the introductory sentence says 'it makes a noise like a pipe when the wind blows') or perhaps we are meant to fill in the details from the very full description of an organ given in the previous chapter. The air pump consists of a piston and cylinder, the piston compressing on the down-stroke (Fig. 6). No valves are mentioned, but we must assume the same kind of arrangement as that given in Chapter 3, except that the cylinder is inverted. It is worked by a rocker-arm, which has on its opposite end a small horizontal plate. The windmill itself is mounted on a separate base, so that it can be turned round as required to face the wind—perhaps through an arc of 90°. It has a single axle, with two discs (*tympania*— 'little drums') on it. One has projecting radial rods which, as it turns round (anti-clockwise in the diagram) push down the small plate and lift the piston. As each rod slips off the plate, the piston is allowed to fall, and its weight then forces the air out into the organ pipes. The other disc on the same axle is fitted with 'vanes, like the so-called *anemouria'.* The word translated as 'vanes' *(platai)* is used elsewhere to mean oar-blades, which suggests that they were wooden, and rigid. The word *anemourion* does not occur anywhere else except as a proper name for a promontory in Asia Minor, but must mean something like 'wind-fan'.

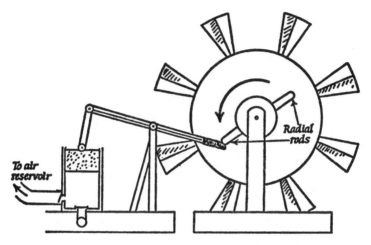

Fig. 6. Hero's windmill for blowing an organ.

Here we have a crude and rather inefficient substitute for the cam, converting the rotary motion of the windmill to the up-to-down motion of the pump. We are not told how many radial rods were used—probably two at the most, since the interval between each thrust would otherwise be too short to allow the piston to empty the cylinder. Hero says 'they (the rods) strike the plate at longish intervals' *(ek dialeimmatos),* which also suggests that the windmill was designed to turn slowly, with a pitch of perhaps only 5–10° on the vanes.

The device was clearly a toy, but why did nobody (apparently) see its potential as a power source? Perhaps its scale was the real reason. A power source, by definition, had to be something which could replace a man or a small animal—that is, something which developed about $\frac{1}{4}$ h.p. at least. It may well be that a small windmill, with rigid wooden vanes, was simply not thought of in this category, and nobody tried experimenting with a bigger and better one. But it is still very puzzling.

STEAM POWER

The failure of the Greeks and Romans to harness steam as a power source was without doubt one of the many factors which prevented industrialization in their society. How near they came to developing a workable steam engine is a much-debated question.

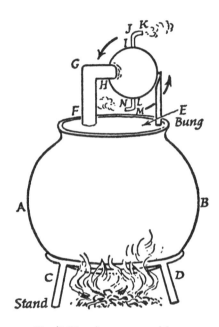

Fig. 7. Hero's steam machine.

Once again it is Hero of Alexandria who provides the only mention in the literary sources of devices worked by steam (*Pneumatica* II, 6 and 11). The second of these is 'a ball which spins round on a pivot when a cauldron is boiled'. He does not give this device a name, though *aeolipyle* (or *aeolipile*) is sometimes used—mistakenly, because that was a different device altogether. The design is simple (Fig. 7). Pressure builds up in the cauldron, and steam passes through the pipe FGH into the sphere, from which it escapes at various points, but mainly through the bent tubes IJK and LMN. As the steam is forced out in one direction (from the outlets), it causes a reaction thrust in the opposite direction, and makes the sphere revolve. The principle is that of jet propulsion, and the device described in Chapter 3 of the same book, in which figurines are made to revolve inside a transparent altar, works in the same way, except that the expansion of heated air is used instead of steam.

Could this form of steam engine ever have been used as a practical power source? The answer is, almost certainly not. It operates best at a high speed, and would have to be geared down in a high ratio. Hero could have managed that, since the worm gear was familiar to him, but not without friction loss. Inadequate heat transfer from the burning fuel to the cauldron would keep the efficiency low, but the worst problem of all is the 'sleeve joint', where the pipe FGH enters the sphere. When making a working reconstruction of this device, I had the greatest difficulty in reaching a compromise between a loose joint which leaks steam and lowers the pressure, and a tight one which wastes energy in friction. It is in the realm of possibility that, given the technology of Hero's age, overall efficiency might have been as low as 1%. If so, then even if a large-scale model could have been built, to deliver $\frac{1}{10}$ h.p. and do the work of one man, its fuel consumption would have been enormous, about 25,000 B.T.U. (26.8×10^6 joules) per hour. The labour required to procure and transport the fuel, stoke the fire and maintain the apparatus would have been much more expensive than that of the one man it might replace, and the machine would be much less versatile.

In his introductory chapter, Hero speaks of his various devices as providing 'some of them useful everyday applications, others quite remarkable effects'. We must conclude that the steam engine came into the second category. Its most remarkable feature is in fact the speed of rotation. My own working model has achieved speeds of the order of 1,500 rpm, and, with the possible exception of a spinning top, the ball on Hero's machine may well have been the most rapidly rotating object in the world of his time.

It is true that this toy (as it may justly be called) does not incorporate the essential elements of a useful steam engine, but it is equally true that all those elements are to be found in various other devices which Hero describes. To make a conventional steam engine it is necessary to develop techniques of making metal cylinders, and pistons to fit them; but this problem was tackled in the design of the force pump, and there is even a possibility that 'lapping' was used (see p. 76). It is not in fact necessary to have an efficient method of converting rectilinear to rotative motion for the construction of a basic steam engine. The earliest working ones were of the beam type, which worked piston pumps without cranks or rotative motion. The one other essential is the valve mechanism, and Hero had devised one for what is usually known as 'Hero's Fountain'—a device exactly like a modern insecticide sprayer, in which liquid is forced out of a container by compressed air *(Pneumatica* I, 10.) Water had to be controlled under pressures of the order of 5–6 lb/sq in (0.35 kg/cm^2). The type of valve he used is not ideal for steam, but it would have done to start with, and might have been improved in the light of experience. An even more significant feature of the 'fountain' is that Hero uses a spherical pressure-vessel on a special stand, showing clearly that he was aware that sheet metal of a given thickness will stand up to greater internal pressure in that form than in any other. In the progress of boiler design, this might have been the first advance on the bunged-up cauldron.

But there was no progress—not even a beginning. What Hero failed to do, and nobody else apparently tried to do, was to combine these essential elements—boiler, valves, pistons and cylinder—to form a steam engine. Why did he not? Perhaps he never thought of reversing the action of a piston pump, forcing liquid or air into the cylinder and taking thrust from the piston. One toy which came very near to this was a 'jumping ball' *(Pneumatica* II, 6), in which a light ball (made of thin sheet metal?) was blown up into the air by a jet of steam from a boiling cauldron. But where expanding hot air, or compressed air, was used to move something mechanically, it was done either by inflating a bladder, which lifted up a weight, or else by shifting water or mercury from one side to the other of a counterbalanced system, which then swung up or down and operated the mechanism by chains and pulleys. These two methods are exemplified in *Pneumatica* I, 38 and 39.

Another explanation which has been suggested is that Hero did attempt to combine the elements of a steam engine, and either blew himself up in the process, or was frightened off the idea. But if the first of these occurred, it is strange that there is no mention of it in ancient tradition. It would have been such a good cautionary tale for anti-materialist philosophers. The second does not ring true psychologically. Stephenson, Diesel and Whittle each persisted with his engine despite nasty accidents and narrow personal escapes. Hero was nothing if not an enthusiastic inventor, and is likely to have thought, as they did, that 'it would never happen to him'.

The other technical reason offered for the failure of this development (as opposed to economic and social factors, which were probably the real causes) was the lack of high-quality fuel. This has implications in a wide range of other contexts, particularly metallurgy. The two fuels in general use were, as one would naturally expect, wood and charcoal (which was called *anthrax* in Greek and *carbo* in Latin). Charcoal was preferred for cooking, because it burned more slowly and made less smoke than wood, and, since artificial draught was used (a fan or bellows), the temperature would be controlled more easily. It was the nearest ancient approach to 'the Heat that Obeys You'. It was also used in metal smelting furnaces. Being almost pure carbon, it is capable under ideal conditions of reaching a very high temperature and, unlike wood, does not contain cellulose or water, both of which slow down the oxidization and limit the rate of heat production.

It is widely held that the Greeks and Romans were unable to get their furnaces much above 1150°C, and that this hindered the development of their iron technology, but recent experiments have suggested that this view is quite mistaken. In a furnace modelled on a Roman type temperatures in the region of 1300°C were reached. If inexperienced modern operators could do this at the first attempt, the skill derived from a year or two of trial and error could surely have produced better results. Such a furnace represents an 'equilibrium situation', in which the temperature rises until the point is reached at which the total heat losses exactly balance the amount of heat being generated. So, to raise the temperature, it is necessary either to increase the heat production, or cut down the heat loss, or both. As far as the first is concerned, little could have been done in the ancient world without the chemical knowledge by which to improve the fuel. On the other hand, charcoal has a calorific value of about 12,000 B.T.U./lb (27.8 × 10^6 joules/kg) and compares favourably with coal. What they could and did do was to cut the heat losses, by designing an effective furnace, by management of the air draught and, above all, by the charging technique—selecting the right proportions of ore and fuel, stacking them in the best way, and replenishing the fuel at the later stages without setting up 'cold spots'. None of this really requires scientific knowledge, merely patience, careful observation and a lot of practice.

In point of cost, charcoal comes out quite well as a fuel. It is dearer than wood, but by how much we cannot really say. Since it is produced by the slow partial combustion of wood, without the addition of any other substances, the difference would be largely accounted for by the labour costs, and it is notoriously difficult to assess the impact of such costs in the ancient world. Not all kinds of wood are suitable. The best are the hardest and most close-grained varieties, such as holm-oak (ilex) and beech. This might add transport costs, since charcoal-burning could only be done in the long term in areas well supplied with such trees, which might be some distance from the main market.

Coal was used as a domestic heating fuel in some parts of the Roman empire (particularly in Britain) but it never made more than a marginal contribution to fuel resources. There is no evidence for deep mining; all the coal used was outcrop, and probably of rather poor quality. It was not normally used in smelting furnaces, though Pliny (*Nat. Hist.* 34, 8, 96) does apparently mention its use in copper casting, which can be done at a considerably lower temperature than iron smelting. In fact, it was not until the invention of coking in the seventeenth century that coal really superseded charcoal as a smelting fuel, and coke stands to coal in much the same relationship as charcoal does to wood—it is much less dense, and has a porous structure which exposes a large area to the air and makes more rapid burning possible.

Charcoal-making was, therefore, a very important activity in the ancient world, but we have very little evidence indeed of the methods used or the degree of competence reached. It was a craft industry, like so many others, and it is most unlikely that any technical manuals were ever written. We do have, however, in a comedy of Aristophanes, a brief glimpse into what might be called (rather pompously) the sociological aspect.

Charcoal-burners in the ancient world, like their few remaining successors today, were normally self-employed. They tended to be individualists of sturdy independence and, because they lived and worked in wooded areas of the countryside, unsophisticated and rather cut off from the current trends of city life. This is partly why Aristophanes chose them to form the chorus of his comedy, *The Acharnians,* named from what was then a small village in 'charcoal country' about 11 km north of Athens. One rather misleading element in their characterization is that they are all very aged, since they claim to have distinguished themselves at the battle of Marathon some 66 years before, which would make them at least 85 or so. But Aristophanes had other very good reasons for making them elderly, which have nothing to do with their occupation, and we should certainly not infer that it was a dying rural craft, confined to the older generation. They are violently nationalist—meaning, in the context, fanatically anti-Spartan—and they attack the envoy who has been sent to negotiate a secret peace treaty. After an unpleasant brush with them (off-stage) the envoy calls them 'Senior citizens of Acharnae, compressed old blocks, ilex-maple-hardwood-veterans of Marathon'. The second of these abusive terms, by the way, probably refers to a process of treading charcoal to make it more dense. When they make their entrance soon afterwards, they live up to this image, since they violently attack the hero, who has dared to make peace terms with The Enemy. He brings about a temporary lull by parodying a scene from tragedy which would have been familiar to the audience; he threatens to 'slay with the sword a hostage he has in his house'. Who can this be? One of their children? It turns out to be a charcoal-burner's pot *(larkos)*, for which they express a deep sentimental affection. 'Spare him !' they cry, 'he is one of us, from our own village!' So the hero escapes, but it was a near thing. Later, when all is explained and forgiven, the chorus sing a little song, ostensibly in their own character, but really in the poet's:

> *Come hither Muse, Acharnian Muse,*
> *Whose song has the vigour of burning fire*
> *Making the sparks fly aloft from holm-oak charcoal*
> *By the fanned breeze agitated.*
> *And the little fish are laid out for the grilling,*
> *And cooks stir up the Tartar sauce*
> *A gleaming garland for the forehead of a lovely sprat.*
> *And others knead the barley-cake;*
> *Come with a rousing rustic song,*
> *Come to me, O Muse—you are one of us!*
> (665–675)

Water supplies and engineering

WATER supply represented one of the most serious problems for Greek and Roman urban communities. With the foundation of each new centre, and each major expansion of an existing one, new supplies had to be found, tested and conveyed—perhaps over some distance—to the delivery point, and at least some storage capacity had to be provided. This might require extensive building works, which had to be planned, inspected and maintained, and inevitably all kinds of legal and administrative problems arose, over both the legitimate rights of consumers, and the illegal activities of mains-tappers and corrupt officials.

It is not surprising, therefore, that Vitruvius devotes a whole book (the eighth) of his *De Architectura* to the subject, and, since this represents a compendium of the theoretical knowledge and practical experience of his day (first century B.C.), it affords a convenient framework for this account. It must be remembered, however, that Vitruvius and Frontinus (the other very valuable source) are writing—from personal knowledge—about the water supply to Rome in the last century B.C. and the following century, which was by far the biggest and most complicated in the ancient world. Some of its features, therefore, may have been unique—for example, the division into three types of supply, and the *vectigal* or water-tax levied on private consumers.

The theoretical basis of hydrostatics, which Vitruvius draws mainly from Greek scientific writers such as Ctesibius and Archimedes, is discussed in Chapter 8. The meteorology (such as it is) comes from even earlier Greek writers. For instance, the account in his second chapter of water evaporation and precipitation differs very little from the kind of popular science expounded by the Sophists in the late fifth century B.C., and parodied by Aristophanes in *The Clouds*. But apart from this general background, there are a few specific principles of physics (or physical chemistry) which are derived from Greek medical writers. One deserves special mention here, and the rest will be discussed as they arise in the course of the account.

At the end of Chapter 1 (and elsewhere) Vitruvius asserts that the influence of sunlight on water is detrimental, since it causes the purest particles to evaporate and scatter, leaving behind the 'heavy, coarse and unhealthy parts'. This is of course perfectly true, and easily observable. One of the tests for purity listed in his fourth chapter—boiling a sample of the water in a metal pot, and seeing whether solid impurities come out of solution—is simply an intensified version of the same process. He does point out, however, that the polluting effect of the sun is most obvious when the water is exposed on level ground, and therefore static. It then causes the growth of algae and insect larvae. The ancients, having no microscopes by which they could see the eggs, believed that such creatures were generated spontaneously by the action of heat on mud. The warmth also causes the production of 'poisonous vapours' —methane and other gases given off by rotting vegetation. All these phenomena (familiar nowadays to amateur water-garden enthusiasts) combine to make water from marshy sites highly unsuitable for drinking, in the unanimous opinion of all ancient authorities.

A curious converse to the 'heat-pollution' idea was expounded by the Greek physician Hippocrates (fifth century B.C.) in his treatise on *Airs, Waters and Places*—namely, that freezing likewise pollutes water, by removing the purest particles, leaving the impurities behind. Though the experiment which he adduces in support (freezing a quantity of water in a metal jar, then thawing it and measuring the loss of volume) is crude and misleading, it is not difficult to see how the idea might have grown up. However clear and pure water is, it becomes opaque when frozen. In the ancient world it must have been very difficult to collect ice or snow (which they used for cooling drinks) without picking up some dust or grit with it. Even today, however carefully one cleans the refrigerator drawer, and however clear the water one puts into it, there always seems to be a spot of sediment at the bottom of the gin-and-tonic when the ice has melted. Though Vitruvius does on occasion speak of water supplies originating from melted snow or ice, be does not mention the 'pollution effect'.

His account of water supplies begins at the basic and practical level, with some procedures for locating underground sources of water. The first and simplest method is to search for water vapour rising from the ground. This is best done at sunrise, when the moisture has risen to the surface (by capillary attraction) during the night, and evaporates as soon as the soil surface is warmed. The best way to observe it, says Vitruvius, is to lie face down on the ground and look along the surface, where the refraction (he calls it 'moisture forming curls and rising into the air') can be most easily seen. This is a sure sign of the presence of water, and justifies a test dig in the area.

Vitruvius then digresses on the types of soil in which water can be found. In ascending order of merit they are (1) clay (limited quantity, bad flavour), (2) loose gravel (water at greater depth, and muddy), (3) peat (water gathers in small droplets, and only collects if there is a sealing basin beneath), (4) gravelly soil and sharp sand, (5) red sandstone rock, provided that the fissures do not drain too much away, and (6) at the foot of a hill, among hard rocks—the best type of source. The only suitable one to be found on an expanse of level ground is an artesian well, with its outlet in a shaded position.

Another good indicator of water resources is plant-life. Vitruvius lists six in particular which will only grow in consistently damp soil—bullrush, wild osier, alder, withy, reeds and ivy. Of course, they usually grow in marshy sites, which are unsuitable as water sources, but where they grow in other situations, and in the suitable soils, they mark a possible source. Needless to say, they must have grown there spontaneously, and must not have been transplanted.

Digging a well is a long and laborious business, and to dig one which turns out to be useless is extremely frustrating. So Vitruvius suggests some additional tests to be carried out before finally starting on the actual digging. If the indications already mentioned are present, a pit should be dug about 1m square and 1.5m deep. A metal basin is to be placed in it, upside down and smeared with olive oil on the inside. The pit should then be covered over with reeds or tree-branches with the leaves still on, and a light covering of soil on top. This should be done in the evening, and the covering left undisturbed overnight. In the early morning it should be opened up and the basin should be examined for droplets of condensation—the olive oil would make these more easily visible. If there is a clear indication of water vapour, it is almost certainly worth while to sink a well-shaft. Other objects may be put into the pit to detect condensation. A unfired clay pot, which is hygroscopic, will go soft (and possibly collapse) during the test. A woollen fleece will not attract so much moisture, but even a small quantity can be detected by wringing it out. Finally, an oil lamp, filled with oil and lit the night before, will not burn completely dry.

Obviously, the ideal source of water is a spring on a hillside, where the water flows continuously through a natural outlet. It can be located without speculative digging, and the height makes it possible to convey the water over some distance by gravity-flow systems. Such sources are not uncommon in Greece and Italy, where limestone formations afford cavities for underground storage. Obvious examples are the Appenine foothills in Italy, and in Greece the range between Attica and Boeotia extending westwards as far as Parnassus, and also the hills of the Argolid and southern Peloponnese. On many of these hills the average rainfall is not much less than that in the drier parts of Britain or many eastern states of the U.S.A., but it is mainly concentrated in the winter months (October–March), which makes storage capacity rather more important than catchment as a limiting factor on supply. The writer of the First Delphic Hymn calls the Castalian spring 'much-watery'. This may sound like a statement of the obvious, but it goes on bubbling up in August after months of drought, and it is easy to see why he took the trouble to mention it.

There are two forms of conduit in which water can be conveyed by gravity flow. The open conduit—much the more common in Roman systems—consists of a channel, usually built into a stone structure and waterproofed with plaster or cement. To keep the level of water even, it has to slope at a more or less consistent angle along its entire length. The gradient was normally between 1 in 150 and 1 in 500 (Vitruvius recommends 'not less than 1 in 200'). A closed conduit usually took the form of a round waterproofed pipe of metal or earthenware, completely filled throughout with water. It can slope up or down at any angle, provided that it does not rise at any point above the level of the intake. Ancient systems on the whole tended to be either all open or all closed, but occasionally a combination of the two forms was used.

The basic problem of the open-channel system was that the gradient had to be maintained consistently over the rises and falls of ground level between the source and the delivery point. If a hill intervened, there were two ways of coping. If there were sufficient masons available, and a copious supply of local stone, a channel was built around the hillside, following the contour line apart from the slight fall required for the flow. It would be supported on what was, in effect, a broad, low wall, with faced stone on the outside and a rubble in-fill, with thin slabs of stone forming the bed and sides of the channel, and a lining of cement to make it waterproof. The Romans called this a *substructio* (Fig. 8).

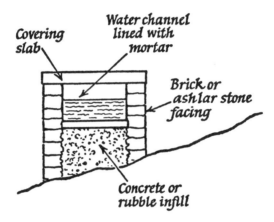

Fig. 8. *Substructio* in Roman aqueduct.

It had three very serious drawbacks. First, it could be very extravagant of materials and labour, since it would have to be built on the bedrock or a very firm subsoil, to avoid the danger of sections being carried down the hillside by heavy rains or landslides. It was very much exposed to pollution, even if covered over with stone slabs—another expensive measure. And finally—a vital consideration for many Greek and Roman cities—in the event of a siege, it was extremely vulnerable to enemy attack. The alternative of tunnelling through the hill was therefore generally preferred.

The usual scheme, as recommended by Vitruvius, was to make the tunnel more or less straight, with vertical shafts up to the surface at intervals of about 116ft (35.5m). This would seem to be a rather excessive provision of shafts, and a number of explanations have been put forward for it. One is that the work-force used on such projects—or at least their supervisors—might have had mining experience, and since the usual ancient mining technique was to sink shafts and join them together with horizontal tunnels ('adits'), this might be a simple transference of the technique from one type of project to another. But the sinking of shafts can be amply justified on two grounds—one applying at the surveying stage, and the other after completion.

It is quite easy to ensure that a shaft is exactly vertical, simply by hanging a plumb-line from a rod across the top, and seeing that the bob hangs in the centre all the way down. If, therefore, a line of posts can be laid (by optical sighting) up over the hill, and shafts sunk from them, the horizontal alignment of the tunnel becomes much easier. Once it has met up with the first shaft, it can be aligned by sighting rods under the centre of each shaft, and will more or less reliably meet up with the next along a straight line. There is evidence to suggest that they did not trouble to get the gradient exactly right at the initial stage, but corrected it later by making a channel in the floor of the tunnel, which could be adjusted a little way up or down as required.

Once the tunnel is made, the air shafts afford easy access to any part of it for inspection and maintenance. An experienced miner could, by regular examination, spot the points at which subsidence or collapse might be expected, and any leakage from the water channel could be promptly stopped. If a major fall took place and a portion of the tunnel became flooded, it would be much easier and safer to locate the exact point of collapse and the extent of the flooding by lowering an observer down a shaft than it would be to send him along the tunnel from the lower end. Finally, the shafts would serve to release pockets of air-pressure which might possibly form if the inflow of water increased very sharply (for instance, after a freak rain-storm) and filled the whole tunnel. Though the shafts might have been expensive in terms of the labour required, they need not have affected the time-schedule very much. The simple fact which makes any tunnelling a slow job is that only a very few men can work on the face at any one time. If all or most of the shafts were dug simultaneously, the work-force could be more efficiently deployed for more of the time.

This type of water-tunnel had been familiar for some time in the Near East before it was first used by the Greeks, and in many modern textbooks (perhaps in recognition of this fact) it is called by the Arabic name *qanat*. The Greeks called it simply an 'excavation' *(orygma)*, and the Romans called the tunnel a 'cave' *(specus)* and the shafts 'wells' *(putei)*. (Fig. 9.)

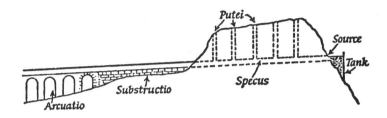

Fig. 9. Roman aqueduct.

One of the most famous Greek examples was built in Samos in the third quarter of the sixth century B.C., when that island was under the rule of the tyrant Polycrates. According to Herodotus (III, 60), the architect was called Eupalinos, and came from Megara, where a similar but smaller tunnel had been built some time before. Substantial remains of Eupalinos' project can still be seen. The tunnel passes through a hill which is almost 1000ft (300m) high. The length is about 1200yds (1100m), and it is roughly square in cross-section, about 8ft (2.5m) in height and width. The point at which the tunnellers, starting from each end, eventually met can be seen, and the smallness of the error in alignment shows how competent the surveying must have been. Remains of other tunnels are few and far between, for obvious reasons. If the channel is not properly maintained and checked, the water over a period of years erodes the walls of the tunnel. If a serious collapse occurs, the upper part of the tunnel becomes flooded, and the water eventually undermines the whole structure. All that remains in many cases is the tell-tale line of holes at regular intervals in the ground above.

Where the ground falls away below the level required to maintain the gradient, the channel must obviously be supported above it. Up to a certain height—around 6½ft (2m) a *substructio*, as described above, was used. At greater heights it becomes more economical to use arches to support the channel, and the very familiar pattern of the aqueduct (in Latin, *arcuatio*) takes over. It should be noted, however, that most of the extended water-channels ran for much the greatest part of their length underground. A typical example is the Aqua Claudia at Rome. Of its total length of more than 35 miles (57 km) only about the last one-seventh was raised on arches.

The construction was usually very simple and straightforward. Either bricks or cut squared stones were used to form the casing of the pillars, with a rubble or concrete filling in the middle. Where the aqueduct passed over a river, the pillars standing in the river bed, and those on adjacent low land if it was subject to flooding, were built with wedge-shaped projections ('cut-waters') to break up the force of the current. The structure containing the channel, which ran above the arches, was usually of the same materials. There was, however, a limit on the height of pillars constructed in this way. It was possible for a very tall one (to put it in the simplest terms) to fold sideways in the middle. This could happen in a very high wind, or if subsidence took place at the base, and if one pillar gave way, it could cause a progressive collapse of the whole series of arches.

The Roman solution to this problem was to limit the height of the arches to about 70ft (21m) or thereabouts, and when they were working near to this limit, they made the pillars very massive, and the arches between them narrow. If a greater elevation was needed, they built the arches in two tiers, the pillars of the upper resting directly on those of the lower. The arches of the lower tier could be made simple and not very heavy, their sole purpose being to brace the pillars from each side. They consisted of the solid, wedge-shaped stones ('voussoirs') forming the arch itself, with shaped stones forming a level top course above the arch. The structure above the upper tier was exactly like that on a single-tier aqueduct. A very good example of the double-tier type survives at Segovia in central Spain, and is still used for part of the town's water supply—a considerable testimony to the skill of the designers and builders. It rises to a maximum height of 164ft (50m) above the ground.

When the aqueduct had to cross a very deep valley, the same principle was carried a stage further. By far the most famous surviving one—the Pont du Gard near Nîmes in southern France—has two tiers of arches, with an additional structure of much smaller arches on top, and the total height above the river bed is

180ft (54.8m). The highest tier is made in the same material as the rest, and carries a water channel about 4½ft (1.36m) wide and 5½ft (1.66m) deep. During the many centuries when water flowed through it, a thick incrustation of calcium carbonate has been deposited on the sides and bottom. Even at this great height, stone slabs were hoisted up and placed over the channel, to shield it from the sun and from pollution.

Ancient water-engineers, especially the Romans, have been subjected to criticism which, being ill-informed, is predictably bitter. It is alleged that they built these elegant and massive structures across valleys unnecessarily, having failed to realize that 'water finds its own level', and that a pipe could have been taken (for instance) down into the valley of the Gard and up the other side. This criticism is wide of the mark on two points. First, it is absolutely clear from the writings of Archimedes, Hero and Vitruvius that they all fully understood the pressure-equilibrium principle, and second, the closed-pipe system is in many ways a much less satisfactory answer to the engineering problems. It is less expensive, but very much more difficult to construct, requiring very specialized skills. It is unreliable, and subject to frequent bursts and leakage. Once constructed, the conduit itself is not accessible for maintenance, and if it becomes blocked, it may have to be completely dismantled and rebuilt. An open-channel aqueduct, by contrast, can be inspected and cleaned regularly, and, as Frontinus points out (II, 124), it is even possible to rig up a temporary by-pass, and carry out repairs on a short section without turning off the main supply.

Two materials are suggested by Vitruvius for the pipes of a closed system—lead (plumbum) and earthenware. He prefers the latter, for several reasons. First, there is a danger of lead poisoning from the formation in lead pipes of white lead oxide (which he calls cerussa)—a highly toxic substance—and as corroborative evidence he points out the unhealthy symptoms shown by workers in lead-smelting and casting. Second, it requires workmen with specialist skills (still called 'plumbers', though they now work in copper and plastics) to carry out construction and maintenance, whereas an ordinary bricklayer can deal with earthenware pipes —or so he says. Third, but not least, lead is a much more expensive material. Nonetheless, he gives detailed instructions for the making of lead pipes.

The Roman method (which can be seen in the remains at Bath in Somerset and in many other sites) was to use a rectangular sheet of lead, which was folded (presumably around a wooden former) into either a circle or a triangle with rounded corners. The two edges either had a simple overlap, or were overlapped and folded (Fig. 10). For special soldering jobs, Hero of Alexandria recom-

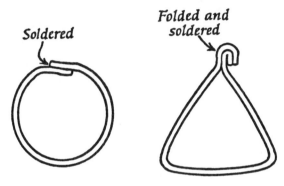

Fig. 10. Roman lead pipes.

mended pure tin, and lead/tin alloys were also used, but if the join along a pipe was made simply by melting the lead or dripping molten lead on to the join, it was probably not very strong or entirely waterproof. The pipes were made in lengths of ten Roman feet, i.e. 9ft 8½in, or 2.95m. There were 10 standard sizes, each named from the width of the sheet of lead used—that is, the circumference plus the overlap, not the diameter as specified nowadays. The sizes are measured in 'digits', this unit being $\frac{1}{16}$th of a Roman foot—0.73in or 1.85cm. From the weight of lead which Vitruvius specifies for one length of pipe, it can be seen that (according to him) the lead sheet ought to be cast, or cast and rolled, to a standard thickness of just under ¼in (0.247in or 6.27mm) regardless of the pipe diameter, which is a little surprising. The sizes in modern units are given in Table 1. The lengths were joined together either by butting them end-to-end and soldering a collar around, or by flaring one end and tapering down the other, inserting the taper into the flare, and sweating the joint together by the application of heat, but how this was done is not clear.

Earthenware pipes *(tubuli fictiles)* were made in shorter lengths,

TABLE OF LEAD PIPE SIZES

Roman size	Lead required per "10-foot" length		Diameter, allowing for overlap ·	
	lb	**kg**	**in**	**cm**
100 – digit	864	392·25	22·6	57·4
80 – digit	691	313·7	17·9	45·5
50 – digit	432	196·1	10·9	27·8
40 – digit	346	157	8·6	22
30 – digit	259	117·6	6·3	16
20 – digit	173	78·5	4	10·2
15 – digit	130	59	2·8	7·2
10 – digit	86	39	1·7	4·3
8 – digit	72	32·7	1·2	3
5 – digit	43	19·5	0·52	1·32

Overlap doubtful, so very approximate in this range

A 20-digit pipe required 1 ton per 125 ft approx.
1 tonne per 37·5 m

Table 1.

but with much thicker walls—Vitruvius recommends 'not less than two digits' (1.46in or 3.7cm). Each section had to be 'tongued' — that is, drawn in to a smaller diameter at one end. This was probably done on a potter's wheel, in which case the length of each section would be limited to about 3–4ft (1–1.2m). To seal the joints, Vitruvius suggests 'quicklime worked up with olive oil'. He concludes with a characteristically Roman tip—crude but practical. When the pipeline is complete, and the water is first let into it, some wood ash should be thrown into the tank at the supply end. This will find its way into any cracks or leaks in the system, and help to clog them up ('grouting' is the technical term). Some modern preparations for curing leaks in car radiators work on exactly the same principle. In the Pergamon water system (to be discussed shortly) some or all of the joints between pipe sections were enclosed in rectangular stone blocks, the pipes passing through a round hole just above the middle of each block.

The two problems with closed-pipe systems are pressure and sediment. If the pipe at any point falls a long way below either the source or the delivery point, the water develops a pressure which works out at approximately 4.4lb/sq in for every 10ft of head, or just under 1 kg/cm^2 for every 10m head. If this pressure rises above the order of 50lb/sq in (3.5kg/cm^2) it begins to have several unpleasant effects. In lead pipes it tends to split open the join, and it will soon find out any flaw or weakness in earthenware pipes. Also, with both types it will tend to blow apart the joints between sections. This is not such a serious problem when they are all in a straight line, or curve gradually up or down, since the joints are held together by the weight of the system as a whole. But Vitruvius points out that if there is a sharp bend between a vertical or near-vertical section and a horizontal one, there is a great danger of bursting, because one of the thrusts (marked B in Fig. 11) has to be taken entirely by the joint itself. The other three (A, C and D) are countered by the weight of the pipes or the supports below the bend. To remedy this, he suggests that when earthenware pipes are used, one should go to the length of enclosing the whole elbow (he calls it a 'knee', *geniculus*) in a block of red sandstone. This sounds rather drastic, but in a '20-digit' pipe (about 4in, 10.16cm in diameter) with a head of 100ft the thrust at B would be of the order of 550lb (250kg), and a cemented pipe joint that could hold together against that would be strong indeed.

This type of closed-pipe system was called a 'stomach' (*koilia* in Greek and *venter* in Latin). In modern textbooks it is usually called a 'U-bend' or (rather perversely) an 'inverted siphon'.

The other problem—sediment—was faced in various ways, the most effective being the provision of settling tanks at the source end. These were long rectangular cisterns of stone or concrete. The water was fed in at one end, and the rate of traverse was slow enough to allow most of the sediment to sink to the bottom by the time it got to the far end. By taking the outlet from high up at that end, near the surface, the purest water was drawn off and fed into the system. Since the tanks had to be drained and cleaned out at regular intervals they were often built in pairs), and used alternately. It has already been pointed out that a very serious problem of maintenance would arise if sediment blocked up the lower part of a U-bend, and a rather cryptic sentence in Vitruvius (the text is almost certainly corrupt) may perhaps give a clue to the Roman answer. In Chapter 6, he says that *colliviaria* should be built into the U-bend, 'to relax the force of the *spiritus*'. *Colliviaria* is a word which is otherwise unknown, and its meaning

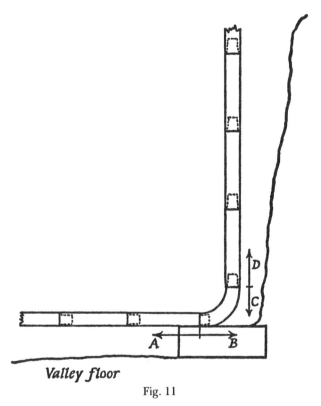

Valley floor

Fig. 11

is obscure. Some editors amend it to *colliciaria*, which is used elsewhere to mean 'gutter-tiles', and interpret the passage as referring to air-valves, used to release air-locks which might form in the pipe. A better emendation, however, is *colluviaria; colluvium* means sludge, so *colluviaria* might well have been sludge-cocks (as they are called nowadays) used to drain the system while repairs are carried out. *Spiritus* can mean air, but in a number of other contexts it is used to mean pressure (equivalent to the Greek *pneuma*), and this meaning would fit exactly here. Vitruvius says that when the system is first completed, the water must be fed in slowly and carefully, or the *spiritus* will burst the pipes. Unfortunately, he gives no details of the design of the *colluviaria*, but they might have been on the lines of the valves used by Hero on his compressed-air fountain (p. 30).

In view of these problems, it is not surprising that closed-pipe systems were exceptional in the Greek and Roman world. But some remains survive of a most impressive one in the Greek city of Pergamon (now Bergama in Turkey) built during the first half of the second century B.C., in the reign of King Eumenes II. According to the archaeologists, the source was high up on a hill above the modern Hagios Georghios. Settling tanks, which probably stood at the inlet end, were built at a height of about 1180ft (360m) above sea level. From there the pipe ran down to a valley about 600ft (183m) below, then rose up about 190ft (58m) over a low hill, down again 130ft (40m) into another valley, and then up again 450ft (137m) to the citadel of Pergamon (Fig. 12). If this was indeed the layout, the pressure at the bottom of the first bend would have been about 260lb/sq in (18.5kg/cm^2).

There is no clear evidence to tell us the material of which the pipeline was made. If it was earthenware, it seems strange that so very little trace survives, as there could be little motive for taking the pipes away, and they could not have been re-used elsewhere. The surviving stone blocks with circular holes (mentioned above) would be appropriate for housing the joints in an earthenware system. If metal had been used, it might well have been removed and melted down, and this would account for the lack of remains. But it is highly doubtful that lead pipes could have withstood the pressure, and if bronze was used (this has been suggested) it must have been very expensive indeed. The total length was over 3 km, and though Pergamon at that time had big resources of money, such a project would have swallowed up a large proportion of the 'gross national product'.

How long this system survived in active use we do not know, but it clearly did not work satisfactorily in the long term. After the city came under Roman government in 133 B.C. it was dismantled and replaced by an open-channel system, carried on arches across the two valleys and through a tunnel in the hill between them. The delivery point was quite low down on the slope of the acropolis,

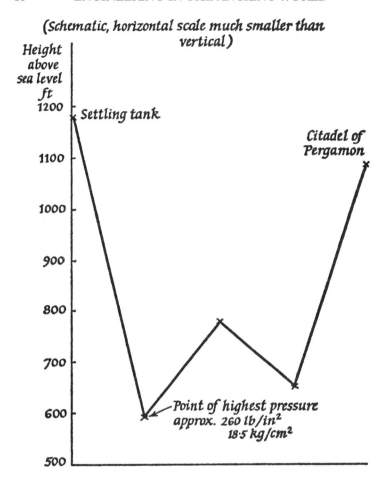

and water for the upper town must have been carried up laboriously, or perhaps pumped up to the supply points.

Material remains of the pipes and accessories by which water was distributed from the supply point to consumers are very scanty, but Vitruvius and Frontinus give us a picture of what they must have been like.

In advanced systems such as the Roman one the supply was divided into three branches, and three reservoirs were built side by side. According to Vitruvius, the central one fed the public supply points (of two types—*lacus,* or pools into which one dipped one's bucket, and *salientes*—water-spouts). These were considered the basic essential, and no other demand was allowed to interfere with them. The reservoir on one side supplied the public baths, and that on the other side the private households who had their own mains supply, and who paid a water-tax *(vectigal)* the proceeds of which were used to maintain the public system. Vitruvius seems to imply that the demand from the central reservoir was constant, whereas that for the baths, and that for private households might vary from time to time. If either of these demands fell off, the surplus which built up in the corresponding reservoir would over-flow into the public supply reservoir, but the levels were so ar-ranged that no amount of extra demand for baths or private cus-tomers could interfere with the public supply.

Frontinus supplies a lot of information on the methods by which supplies were measured and assessed for tax. Here, as in so many contexts, we meet the characteristic contrast between knowledge and understanding of the static features, and neglect or ignorance of the dynamic. No attempt seems to have been made to measure the speed of flow through a pipe or conduit, and the whole technique of measurement is based on a calcula-tion of the cross-section area of a special nozzle which regulated the flow. It is clearly recognized that if the gradient of an aque-duct is steeper, the rate of flow will be faster, but Frontinus makes no attempt to find out how much faster. He is concerned merely with a nozzle of a specified size. If the rate of flow is normal, such a nozzle will deliver the statutory amount of water to the consumer. If the gradient of the channel is extra steep, or if the supply is increased by heavy rainfall in the catchment area, the amount delivered will be above the legal requirement *(exuberare),* but nothing is done about this—it is simply regarded as a bonus to the consumer. He does speak of making some adjustment if the rate of flow is slower than normal, but exactly what he means by 'lightening' *(relevandam)* is not clear.

The nozzle (in Latin, *calix*) which regulated the supply to a private consumer was made of bronze, this being a harder metal than lead, and therefore less easily tampered with. It was about 9in (22–23cm) long, its internal diameter carefully measured, and an official tester's stamp on the outside—at least, it should have had one. It was set in the wall of a reservoir as a rule. It could be let into a pipe or conduit, but this was avoided where possible, as it was liable to cause leaks. The supply pipe was attached to the outlet of the nozzle. Frontinus notes that the position of the nozzle affected the amount drawn off. If set at an angle 'facing the flow' (Fig. 13) it would obviously collect more, and if slanted the other way, less. Also, the level of the pipe leading away from the *calix*

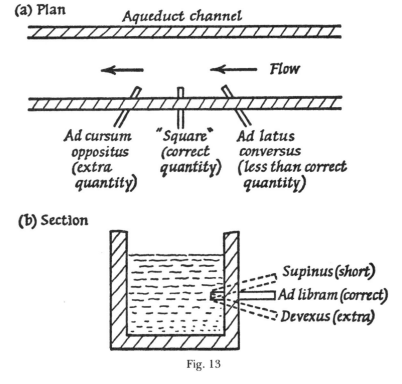

(a) Plan

Aqueduct channel

Flow

| Ad cursum oppositus (extra quantity) | "Square" (correct quantity) | Ad latus conversus (less than correct quantity) |

(b) Section

Supinus (short)
Ad libram (correct)
Devexus (extra)

Fig. 13

affected the rate of flow. If it sloped downwards steeply *(devexus)* the flow would be increased. And though Frontinus requires that the nozzle should be set at right-angles to the flow, and that the pipe leading from it must be level for some distance, he does not stipulate a particular depth (e.g. x digits below the surface), which would have given some kind of consistency to the measurement. It should be said, however, that the system was fair to the extent that if the supply was a bit short (owing to drought) the loss to each customer would be proportionately the same, or roughly so.

Theoretically there were 24 different sizes of *calix*, but in practice only 15 were used, and the most important ones are listed in the Appendix to this chapter. The smallest was called a *quinaria*, 'five-to-each-customer', and had a diameter of about $\frac{9}{10}$ in (2.31 cm). This was used as the standard unit of measurement for the larger nozzle sizes.

For the sake of brevity, let us call this a 'no. 5'. There were six others between that and the 'no. 20', namely nos. 6, 7, 8, 10, 12 and 15. The 'no. 20' was about $3\frac{1}{4}$in (8.316cm) in diameter, and was the standard supply size for a small group of customers. Above that, there were seven larger sizes in general use, nos. 30, 40, 50, 60,70,80 and 90, and the two 'supply main' sizes were the no.100 *(centenaria)* and no. 120 *(centenum vicenum)*. The sizes of the smaller and larger nozzles were worked out on two different principles. Up to the 'no. 20' the number reflected exactly the internal diameter expressed in quarters of a digit (the digit being $\frac{1}{16}$ of a Roman foot, 0.7275 in or 1.848 cm). Thus the 'no. 5' was $\frac{5}{4}$ digits diameter, the 'no. 6' $\frac{6}{4}$ (i.e. $1\frac{1}{2}$) digits, 'no. 8' 2 digits, and so on. For the 'no. 20' and above a different formula applied. The number denoted the cross-section area of the nozzle expressed in square digits. Thus the 'no. 40' was 40 square digits (136.56cm^2) and the 'no. 100' 100 square digits (341.4cm^2). As Frontinus points out, the two principles of measurement coincide (nearly but not exactly) at the 'no. 20'. In addition to these complications, the men in charge of the supply connection and maintenance ('watermen', *aquarii*) had other resources with which to confuse and cheat their customers. In some parts of Italy the standard unit was apparently the inch (*uncia*, $\frac{1}{12}$ of a Roman foot, 0.97in or 2.464 cm) instead of the digit. Again, it was not too difficult to create confusion between a nozzle of $\frac{5}{4}$ digits diameter with one of $\frac{5}{4}$ square digits area. Even within a highly organized system, subject to inspection and control, certain dishonest manipulations were practised. According to Frontinus, in many situations water was supplied to a reservoir through a 'no. 100' or a 'no. 120', and the distribution pipes (for each group of— say—a dozen customers) were 'no. 20'. But the *aquarii* used a 'no. 100' which was about 13.5% bigger than it should have been, and a 'no. 20' about 20% smaller than the correct theoretical size. They thus had a surplus amount in the reservoir, and were able to supply some extra customers illicitly, pocketing the water tax ('You pays the money to me, guv'nor, and I squares it up at 'ead office') or at least a substantial bribe. Since the unofficial customers (again according to Frontinus) included the proprietors of bawdy-houses, one can see that the career of an *aquarius* was beset with manifold temptations.

There is one simple and obvious question which, surprisingly, cannot be answered directly from the evidence available. If a Roman householder had a piped supply of water, did he have a tap to turn it on and off? The fact that neither Vitruvius nor Frontinus makes any mention of such a device may merely mean that they regarded it as too familiar to need description. The suggestion mentioned earlier, that the rate of consumption from the private supply reservoir might vary, cannot prove anything either. It might simply mean that more or fewer customers might be connected up at any one time. If there were no taps, the water presumably ran from a spout into a basin, from which it flowed away (perhaps being used to flush a lavatory en route) into one of the drainage channels, and thence into the Tiber. This was certainly the arrangement at the public water supply points.

All this gives the impression that the water supply to Rome was copious, and indeed, by any European standards up to the nineteenth century, it was very copious. Exact calculations are impossible, because the measurements Frontinus gives are of the cross-section area of the various aqueduct channels, and even after Thomas Ashby's meticulous examination of the remains it is impossible to determine the gradients, and hence make a rough guess at the velocity. However, a conservative estimate comes out at about 150–200 million gallons (680,000–900,000 m^3) per day.

This, however, represents the heyday of water engineering in the capital city of the Empire, which had financial resources and expertise available in full measure. In the provinces, small communities had less happy experiences of what was, to them, advanced technology. A graphic record of one fiasco survives in a stone inscription* found in the Roman town of Saldae (now Bougie in Algeria, on the coast about 120 miles—200km—east of Algiers).

* *Corpus Inscriptionum Latinarum* VIII, 2728.

It was set up in 152 A.D. by Nonius Datus, a retired army surveyor *(librator)* to celebrate his own achievement in rescuing an important project from disaster. After telling how he was summoned, at the Emperor's personal request, to take over control of the work, he says: 'I set out on the journey, and was attacked by brigands; I and my staff were robbed of everything, and badly injured. I came to Saldae and met Clemens the Provincial Governor, who took me to the hill, where they were shaking their heads and weeping over the tunnel, on the point of giving up the whole thing. The tunnellers had covered a distance greater than that from side to side of the hill!' He explains that the inlet end of the tunnel, running eastwards, had gone off alignment to the right (i.e. southwards), and the tunnellers who had started from the other side had done the same, turning northwards. So, naturally, they had not met in the middle as planned. Nonius Datus did a fresh survey, much more meticulously this time, and by encouraging competition between the two sets of tunnellers, got the work completed in reasonable time. The inscription ends with a reference to the great local occasion when the supply was ceremonially turned on by the Provincial Governor.

Appendix to Chapter 2

THE SIZES OF MEASUREMENT NOZZLES, AND FRONTINUS' ARITHMETIC

The section of Frontinus' work which deals with nozzle sizes (I, 23–63) affords us an opportunity to judge the skill and accuracy in arithmetic shown by an educated but not technically minded Roman administrator. (On Frontinus' career, see Chapter 9.)

For his calculations he uses a system of fractions. In some places these are written out, like the whole numbers, in words, while in the last section (39 onwards) he uses the normal Roman notation for the whole numbers, and a system of symbols for the fractions.

The system is duodecimal, and the basic fraction is the *uncia*, $\frac{1}{12}$, indicated by a dash,—.

Thus = means $\frac{2}{12}$ ($\frac{1}{6}$),

= = − means $\frac{5}{12}$, and so on

S (semis) is used for $\frac{1}{2}$,

so S = − means $\frac{1}{2} + \frac{3}{12} = \frac{3}{4}$, and so on.

Fractions of a twelfth in general use are the half ($\frac{1}{24}$), written S or £, the sixth (*sextula*, $\frac{1}{72}$) for which there is no symbol, and the twenty-fourth ($\frac{1}{288}$) called a *scripulus* and written ^{3}. (This is the 'scruple' of the old Apothecaries' Weight, one twenty-fourth of an 'ounce', which was one-twelfth of a pound.) For extra accuracy, Frontinus occasionally subdivides the *scripulus* into thirds ($\frac{1}{864}$). Apart from these basic fractions more or less any numerator or denominator can be used, but they have to be written out in words—there are no symbols.

In calculating circumferences and areas of circles, Frontinus takes the value of π as $\frac{22}{7}$ (3.1428571), which is about 0.04% higher than the true value. Archimedes had shown three centuries earlier (in his treatise *The Measurement of the Circle*) that the true value lies between $\frac{22}{7}$ and $\frac{223}{71}$. The latter is in fact slightly closer to the true value (−0.024%), but Archimedes did not establish that fact, and $\frac{223}{71}$ would obviously be much more awkward to use as a formula.

Making allowance for this inaccurate value of π, Frontinus' arithmetic is quite creditable by any ordinary standards. He says, for instance (I, 24), that the difference in area between a square digit and a circle of 1 digit diameter is $\frac{3}{14}$ of the former and $\frac{3}{11}$ of the latter. Taking the digit as 1.848cm, 1 square digit is 3.415104cm^2. The radius of the circle would be 0.924cm, and πr^2 (taking $\pi = \frac{22}{7}$) would be 2.683 296. The difference is

$$
\begin{array}{r}
3.415104 \\
-\ 2.683296 \\
\hline
= 0.731808
\end{array}
$$

$3.415104 \times \frac{3}{14}$ = 0.731808

$2.683296 \times \frac{3}{11}$ = 0.731808

So the fractions $\frac{3}{14}$ (*tribus quartisdecumis suis*) and $\frac{3}{11}$ (tribus *undecumis suis*) though they sound like rough approximations, are in fact accurate to at lease six places of decimals.

Table 2. The most common sizes of *calix* listed by Frontinus.

Number	Latin name	Diameter		Circumference		Area		Capacity
		digits	cm	digits	cm	sq. digits	cm²	quinariae
5	quinaria	$1\frac{1}{4}$	2.31	$3+\frac{267}{288}$	7.26	—	4.191	1
8	octonaria	2	3.696	$6+\frac{82}{288}$	11.611	—	10.728	$2+\frac{161}{288}$
12A	duodenaria	3	5.544	$9+\frac{123}{288}$	17.417	—	24.14	$5+\frac{219}{288}$
12B	duodenaria	$3+\frac{18}{288}$	5.659	—	17.779	—	25.151	6
20A	vicenaria	$5+\frac{13}{288}$*	9.323	$15+\frac{246}{288}$	29.29	(20)	68.265	$16+\frac{7}{24}$
20B	vicenaria	$4\frac{1}{2}$	8.316	—	26.125	—	54.315	13
40	quadragenaria	$7+\frac{39}{288}$	13.186	$22+\frac{5}{12}$	41.425	(40)	136.56	$32+\frac{7}{12}$
60	sexagenaria	$8+\frac{212}{288}$	16.144	$27+\frac{17}{24}$	50.71	(60)	204.69	$48+\frac{251}{288}$
80	octogenaria	$10+\frac{26}{288}$	18.646	$31+\frac{17}{24}$	58.58	(80)	273.06	$65+\frac{1}{6}$
100A	centenaria	$11+\frac{81}{288}$	20.847	$35+\frac{11}{24}$	65.495	(100)	341.33	$81+\frac{130}{288}$
100B	centenaria	12	22.176	—	69.668	—	386.24	92
120A	centenum-vicenum	$12+\frac{102}{288}$	22.83	$38+\frac{1}{12}$	71.724	(120)	409.35	$97\frac{3}{4}$
120B	centenum-vicenum	16	29.568	—	92.89	—	686.64	$163+\frac{11}{12}$

* By the system used for the smaller sizes, this should be exactly 5 digits ($20\times\frac{1}{4}$), but Frontinus has corrected it to bring it into line with the other system, so that the cross-section area is 20 square digits.

As stated in the text, Frontinus takes $\pi=\frac{22}{7}$, but in calculating the centimetre equivalents, the value 3.1416 has been used.

Where two sets of measurements are given, the first in each case (12A, 20A, etc.) is the theoretical value calculated by Frontinus, and the second (12B, 20B, etc.) the values used by the *aquarii*.

The most common sizes of *calix,* as calculated by Frontinus in I 39–83, are listed in Table 2. For the sizes up to no. 20 he starts from the diameter in quarter-digits, as explained on p. 51. From this he works out the circumference (col. 3) and the capacity in *quinariae* (col. 7), this being the ratio of the cross-section area to that of a *quinaria* nozzle, $1\frac{1}{4}$ digits diameter (4.191cm²). A spot check on a few of his figures, chosen at random, gives an indication of his accuracy.

Take his figures for the 'no. 12A' nozzle. The diameter is 3 digits (5.544cm). He gives the circumference as $9 + \frac{123}{288}$ digits (9.4270833) and the true value is 9.4285714. The difference is 0.001488 1 (or, in his terms, less than one-third of a *scripulus*) and he error is approximately –.015%. The capacity in *quinariae* he gives as $5 + \frac{219}{288}$ (5.7604166), while the correct figure 5.759 999 5. The error is 0.000 417 1 of a *quinaria,* or +0.007 2%.

For the sizes above no. 20 he starts from the figure in col. 5 and works back to col. 1, which involves extracting a square root, e.g. for the 'no. 80', d=2 $\sqrt{\frac{80}{\pi}}$. His answer, $10 + \frac{26}{288}$ (10.090 277) requires something better than 4 figure tables to correct it. Checking back from that figure to column 5 (still taking $\pi = \frac{22}{7}$) gives a figure of 79.996 468 square digits, an error of –0.004 4%, which has been compounded by squaring. From the figure in col. 1 the circumference works out at 31.712 298; Frontinus gives $31 + \frac{17}{24}$ (31.708 333), an error of –0.012 5%. The figure for *quinariae* in col. 7 is $65 + \frac{1}{6}$ (65.166 6), and error of 0.003 of a *quinaria,* or –0.0046%.

The exceptions to this generally good standard of accuracy are the figures given for sizes 'as used by the watermen'—12B, 20B, l00B and 120B. It may be significant that no figures are given for the circumferences (this being the crucial measurement in the manufacture of the *calix),* and the other figures, with two exceptions, are all in round numbers ($4\frac{1}{2}$, 12 and 16 digits, 6, 13 and 92 *quinariae).* The remaining two figures can be reasonably explained. The 'watermen's no. 12' was supposed to deliver 6 *quinariae,* as against the $5 + \frac{219}{288}$ of Frontinus' calculation (no. 12A), and accordingly, the diameter is corrected by $\frac{18}{288}$, which is within a very narrow margin of the correct amount. The *quinariae* for the 'watermens' no. 120' present a more difficult problem. It looks as though Frontinus has worked it out properly, instead of accepting a round figure (the watermen probably reckoned it as 164), but he goes on to say that it is twice the capacity of a no. 100—that is, his own theoretical 100A. In this he is a whole *quinaria* out, and unless one assumes some alteration in the text, it must be admitted that here, for once, he has been a little careless.

3

Water pumps

In the ancient world, pumps and similar devices were used for various purposes. Much the most important was the irrigation of crops, especially in market-gardens, but they were also used for pumping flood water out of mines, bilge-water out of ships (the larger and better equipped merchantmen) and, according to Hero of Alexandria, in fire-fighting equipment.

The simplest and crudest dates from the very remote past, and was in use all over the Middle East long before the classical period, and illustrations of it are found on Greek vases of the sixth century B.C.* from the time when work scenes first begin to appear as subjects in vase-paintings. Nowadays it is usually called a swing-beam or swipe, or by its Arabic name *shadouf.* The Greek name for it was *kelōneion,* and the Latin names *ciconia* ('stork', from its resemblance to that bird) and *tolleno* ('lifter'). It consists of a support, usually a tree-branch with a fork at the top, driven into the ground about 8–10ft (3m) away from the well-head or other source. This acts as a pivot for a beam, one end of which is directly above the well-head, and has a bucket suspended from it on a rope. Towards the other end is a counterweight, commonly a large stone, which is shifted along the beam until it just out-weighs the bucket full of water. The rope is pulled down until the bucket dips into the water and fills. It is then released, and the counterweight lifts the bucket a few feet above the ground, where it is emptied by hand into another receptacle, or into a conduit of some sort.

Using one of these machines is not much less laborious than using a bucket on a rope—it is a little easier to reach up and pull downwards than to reach down and pull upwards. In an amusing passage in Menander's comedy, *The Misanthrope (Dyskolos)* a wealthy

*e.g. Attic black-figure vase in Berlin, reproduced in E. Pfuhl, *Malerei und Zeichnung der Griechen* (Munich, 1923), vol. III, pl. 72, no. 276.

young man who, for romantic reasons, has been using a mattock, describes how his back, shoulders and hips all locked solid from the unaccustomed hard work, until he was 'swinging up and down in one piece like a shadouf''. It was clearly a familiar sight to the Greeks and Romans, but was so simple and obvious that the writers on machinery did not consider it worth mentioning.

In many situations, water for irrigation had to be raised from a stream or canal, over a low bank and onto an adjoining field. The quantity required was large, but the head lift only a few feet. For this purpose two types of pump were used, the screw and the drum. The Greek word *mēchanē* ('machine') is frequency used in contexts connected with irrigation, and might refer to either type, and it is worth noting that the need for such a device was so widespread in Egypt that in Greek documents from Oxyrhynchus a piece of cultivated land comes to be called 'a *mēchanē*'.

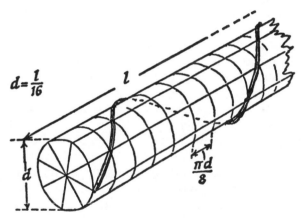

Fig. 14. Archimedean screw pump.

Where a screw pump is specified, it is called *cochlias* in Greek and *cochlea* in Latin, from its resemblance to a spiral sea-shell. Tradition ascribed its invention to Archimedes in the third century B.C., and though this may be true, it cannot be relied upon absolutely—the pump might have been in use earlier. Vitruvius gives a detailed description of how it was made (X, 6). The rotor is made from a round wooden pole, its diameter $\frac{1}{16}$th of its length (Fig. 14). The circumference is divided into eight equal segments, marked by parallel lines along its length. Then the length is divided into sections, each equal to one-eighth of the circumference, and marked by rings drawn around the circumference. The effect of this is to produce a pattern of small squares all over the surface, and since the blades are placed diagonally across these squares, the rake angle of the spiral is 45°.

The blades are constructed by taking a flat strip of willow or osier, coating it with pitch and placing its end on one of the longitudinal lines at one end of the shaft. Then it is drawn obliquely around the rotor shaft and nailed onto it at the intersection of the two lines, ⅛th of the circumference along and the same distance around the shaft. It is then taken along the same diagonal course and fixed over the series of intersections, returning at the eighth point to the longitudinal line from which it started, having made one complete circuit of the shaft. Then it continues along the same path to the far end of the shaft, making just over five circuits altogether. This forms the base of one blade, and more strips are pitched and nailed on top of it until the diameter of the rotor is built up to twice the original diameter of the shaft (i.e. ⅛th of its length). Vitruvius' account suggests that seven more blades were then positioned by the same method, one starting from each of the longitudinal lines. This is by no means impossible, but it raises some problems, which will be dealt with later.

A wooden cylindrical case is then made to fit around the blades (however many), constructed like a barrel, the planks painted with pitch and bound with iron hoops. There is virtually no doubt that in Vitruvius' design the case was fixed to the rotor blades. There is some obscurity in his description of the way in which the rotor was mounted, but it is clear that it had an iron cap on each end and (probably) an iron spigot which turned in an iron socket on each of the supporting posts. The rotor with its case was turned 'by men treading' *(hominibus calcantibus)* but exactly how is not clear. Vitruvius recommends that the rotor should slope upwards at an angle of about 37°. Though this is arrived at from the well-known construction of a right-angled triangle with its sides in the ratio 5:4:3, it probably represents an approximation to an optimum angle which has been arrived at empirically. ('Us put 'un up like this-yur, an' 'ee' wurked allright; us put 'un higher an' 'ee didden wurk so gude, so us put 'un back where 'ee be, an' let 'un bide'.)

How, then, was it turned? Vitruvius provided a diagram in his original work, but this has been lost, and the drawings which appear in our earliest manuscripts of the *De Architectura* are the work of medieval scribes and commentators. Most of them show a treadmill mounted outside the case (clearly, they assumed that it was fixed to the rotor) rather improbably placed halfway down and tilted at the same angle of 37°. Even if it were in a more convenient position, it would still be extremely difficult to tread a wheel tilted at such an angle and, needless to say, very inefficient. A terracotta in the British Museum* shows a workman steadying himself on a crossbar at about waist height, and turning a cylindrical object with his feet by treading a row of cleats around its middle. The fact that the cylinder tilts upwards towards the right suggests that it might be a screw pump, but the tilt is very much less than the angle Vitruvius suggests, and the modelling is crude and perhaps inaccurate. Another ancient illustration is to be found in a wall-painting at Pompeii†. This shows a slave standing in the shade of a small structure like a lych-gate, turning a compact-looking cylinder with his feet. It is generally taken to be a screw pump, but this may not be correct. The output is pouring into what looks like a storage jar buried in the ground, so it might possibly be a mill or crusher of some kind. The angle of tilt is, if anything, even less than in the terracotta, and the lift appears to be no more than a few inches.

Some remains of Roman screw pumps and their mountings have been found in Spanish mines, and they appear to have been tilted at an angle of about 15°. Thus there is evidence for a variety of different inclinations, and three comments should be made. Firstly, the best angle for efficiency is clearly related to the rake angle of the spiral, and this was no doubt taken into account when designing the screw. Secondly, for a given rake angle of screw, the amount of water lifted by each blade is reduced if the pump is tilted to a higher angle, and eventually drops to zero when the tilt becomes equal to the rake angle. Vitruvius' pump would cease to work altogether above 45°, and the angle he suggests (37°) was apparently chosen to give the maximum possible lift at the cost of reduced output. A different compromise would be reached if

*3rd–2nd Cent. B.C. Illustrated in Henry Hodges, *Technology in the Ancient World* (Pelican, 1971 fig. 211, p. 184).

†In the Casa dell' Efebo. Illustrated in Rostovzeff, *Social & Economic History of the Roman Empire* (second ed. O.U.P. 1957) pl. LIII, 5 and R. J. Forbes, *Studies in Ancient Technology* (Leiden, 1963) vol VII, p. 213.

output were more important, and lift mattered less. Incidentally, the device was, very roughly speaking, self-compensating. As increased tilt increased the head but decreased the output, the work done at any angle, and hence the power required, would remain approximately constant for a given speed of rotation. Thirdly, the number of blades that could be mounted on the rotor might be limited by the angle of tilt. If the head is low, the quantity of water lifted by each blade may be curtailed if the next blade above it is too close. Here again, Vitruvius' design is a reasonable compromise. If, for instance, the inside diameter of the pump case was about 1ft (30cm), the water above each blade would extend about 7.4cm up the inside of the casing. With all eight blades mounted on the shaft (as Vitruvius seems to suggest) the spacing between them would be about 5.9cm, and only a minimal reduction in the amount of water lifted by each blade would result.

Surviving remains of rotors from screw-pumps do not apparently have as many as eight blades. One of the most interesting, in the Liverpool Institute of Archaeology,* has two, starting from diametrically opposite points, and the rake is about 45°. One explanation might be that it was intended to work at a lower angle of tilt. The addition of extra blades would, of course, increase the output per revolution, but it would add appreciably to the weight and the cost of the pump, and would be pointless if the available driving power could be fully taken up with fewer blades. Finally, one very significant advance in design is evidenced by a surviving screw found in the Centenillo mines near Linares. Its blades were made from sheet copper, ⅛in (3.2mm) in thickness, and were fixed to the rotor and to the case by small metal brackets.

To return to the question of the angle of tilt, there is one more observation which may be relevant. Vitruvius describes, as though it were something not particularly remarkable, a right-angle drive in the water-mill. It is quite clear from this that the engineers of his day were capable of devising gears which could have coupled a horizontal treadmill shaft to a tilted pump shaft, but there is no evidence to show that this was actually done in antiquity. Whatever the arrangement may have been, the man or men turning the pump used a treading action, and it is an interesting commentary on human progress that the photographs of screw pumps in

*Illustrated in T. A. Rickard, *The Mining of the Romans in Spain*, J.R.S. VIII (1928) 129–143, and Plate XII/1.

use in Egypt today almost all show two men seated on the ground, one holding the mounting steady and the other turning the rotor with a handle. As anyone knows who has cranked a reluctant car engine on a winter morning this is a much less efficient and much more back-breaking way of harnessing human power.

It would be most interesting to reconstruct a pump of this kind to Vitruvius' specification, test it at various angles of tilt, and make an assessment of its efficiency. But the best that can be offered here is a very approximate guess at its performance, based on a number of assumptions which are arbitrary, and may prove to have been incorrect. Suppose the rotor to be nearly 8ft (2.4m) in length, and nearly 1ft (0.3m) in diameter. If mounted as Vitruvius suggests this would raise the water to a head of nearly 4ft (1.16m). Given an available power, from one 'man treading', of 0.1h.p. the amounts pumped at various rates of efficiency would be as follows:

at 60% efficiency just over 50 gall (about 235*l*) per minute
at 50%   "   just over 40 gall (nearly 200*l*)   "
at 40%   "   nearly 35 gall (nearly 160*l*)   "

There would be appreciable loss of energy through friction in the rotor shaft bearings, and also through spillage of water over the rotor shaft due to uneven motion of the rotor, but it would be reasonable to guess that the efficiency might be somewhere in the region of 40–50%. Even so, the output is by no means contemptible. 2,000 gallons of water in an hour could make a vast difference to a parched vegetable garden.

The other type of low-lift, high output pump described by Vitruvius (X, 4) was called *tympanon* in Greek (spelt *tympanum* in Latin), the name meaning 'drum', and is simpler to construct than the *cochlea*. It has a horizontal axle, turning in bearings supported on posts or beams at each end. Vitruvius recommends that both the axle and the bearings should be 'plated with iron', but this may not have been the usual practice. A few Greek papyrus documents from Egypt refer to the supply and collection of replacement wooden axles (and, incidentally, the 'handing-in' of the old ones), from which it is clear that wear on the axle was a major problem. On this axle turned the drum itself, made from planks in the form of two parallel discs, joined together by eight partitions which divided the drum radially into eight

compartments (Fig. 15). Hence the pump is sometimes referred
to as a 'compartment wheel'. The sides and partitions were all
sealed with pitch 'like a ship's hull', and the outside rim was also
boarded in, except for a slot about 6in (15cm) wide, opening into
each compartment at the end which was to enter and leave the
water first as the drum revolved. A circle of eight holes was cut in
one side of the drum near the axle, one opening into each compartment, to act as outlets for the pump, and a wooden trough (known

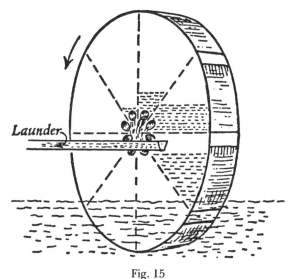

Fig. 15

as a 'launder') was placed so as to catch most of the water as it
flowed out, and channel it into a conduit. The drum, like the screw
pump, was turned by 'men treading'. If the drum itself was made
to act as a treadmill by fixing cleats around its rim, care would
have to be taken to avoid excessive resistance as the cleats passed
under water. There may have been a separate treadmill mounted
on the same axle, though Vitruvius does not specifically say so. In
any case, the axle was horizontal, and the problem of tilting did
not arise.

The design of this pump is very simple and reliable, and the
only loss of energy (though it might be serious) occurs in the axle
bearings and in the 'paddle effect' of the cleats passing through
the water. But it has a number of limitations. Firstly, the absolute

maximum height to which it can raise water is a little less than half its own diameter, and in order to raise a significant quantity, the head must be reduced to something like two-thirds of that height. This means, for instance, that a drum nearly 10ft (3m) in diameter would not have a very effective output at a head of more than 39in (1m). If a drum of this diameter had a thickness, measured internally, of nearly 8in (20cm) each compartment would scoop up about 11 gall (50*l*) of water, giving an output of about 88 gall (400*l*) per revolution. A very approximate idea of the pump in action can be obtained by supposing that the drum revolved at 2 rpm, which would produce an output of about 176 gall (800*l*) per minute. To turn it at this speed would require at least two men working quite hard (0.2 h.p., with 20% friction loss) and it might not be possible for them to keep it up over a long period. To make the picture more precise, let us imagine the drum itself forming the treadmill, with cleats around its circumference at intervals of just under 10in (25cm). There would be 38 treads altogether, and to turn the wheel at 2rpm the men would have to tread 76 times per minute. It would feel like climbing a steep flight of stairs at a brisk walking pace.

In addition to the limitation on the head lift, the drum has two others which the screw pump does not have to the same extent. One is that the output has to flow through the holes near the axle. No dimensions for these are given by Vitruvius, but his term for them, *columbaria* ('dovecote') suggests that they were not large, and their width is in any case limited by the tapering of the compartments towards the centre. As a result, there is a limitation on the output of the pump which is reached when the speed of revolution leaves insufficient time for the compartments to empty as they pass over the axle. Beyond that point increasing the applied power, and the speed of rotation, will not bring about a proportionate increase in output. But with the screw pump the speed and output can be increased indefinitely—until, that is, the casing bursts or the lower support disappears into the mud. It is unlikely, however, that there would ever be sufficient power available to turn the drum fast enough to encounter this limitation, except in unusual emergencies, such as sudden flooding of a mine-shaft, or desperate renewal of supplies for irrigation after a long drought.

The third limitation must have been encountered quite often when raising water from a muddy ditch or canal. There is a tendency for mud and debris to pack around the outlet holes and reduce the flow through them. Because the rotor of the screw pump presents, in effect, a smooth channel without sudden bends, it is capable of pumping slurry with quite a considerable solid content—though there comes a time, of course, when it eventually clogs up. This is why the screw pump is still in use nowadays for such purposes as pumping untreated sewage, or clearing animal refuse from farm buildings. It is usually called an 'auger', and is made with a fixed case and a steel rotor.

In conclusion, a word should be said about the comparative efficiency of the screw pump and the drum. In the figures for performance given above—which, it cannot be too often stressed, are mere guesswork—it has been assumed that the drum was rather more efficient than the screw. As against this, ancient writers in more than one context comment on the remarkable efficiency of the screw. Two examples will suffice. Diodorus Siculus (V, 37, 3–4) remarks that a surprising quantity of flood water was pumped out of Spanish mines by this means, and in a description of a merchant ship built in the third century B.C. (Athenaeus V, 208 F) which was enormous by the standards of the time, it is explicitly stated that one man was able to do all the bilge-pumping required by means of a screw pump. It must be said, however, that this claim should be regarded with suspicion. From a rough estimate of the overall size (based on the cargo capacity) we may infer that the ship's draught was in the region of 15ft (4.6m), and a single screw-pump would hardly have given the necessary lift.

Where it was necessary to raise water to a greater height, a number of other devices were used. All of them suffered from the drawback of a limited output, and where both a high lift and a large output were required, the only course open was to construct a series of pumps feeding each other. This must have been very costly to build and operate, and only in a situation such as silver mining, where increased output of ore might offset the costs, would it have been worth while. In agriculture this would hardly ever be so.

The simplest of the higher-head devices was the bucket-wheel, a modification (or possibly a precursor) of the drum, described very sketchily by Vitruvius later in the same chapter (X, 4). The Greek name for it was *polykadia* ('multi-bucket'), but Vitruvius does not give a Latin name. It consists of a wooden wheel with buckets fixed around its rim, and its axle mounted at such a height as to allow the buckets to dip just below the surface of the water to be lifted as the wheel is turned. Each bucket scoops up a quantity of water, and as it passes over the top of its orbit its contents are tipped into a wooden trough The shape and positioning of the buckets is clearly very important. Vitruvius calls them *modioli quadrati* (rectangular grain-measures), which perhaps indicates that they were broad at the base and tapered to a narrow opening at the top since this was the usual shape for such a measure. He also says, 'when they have reached the high point and are returning downwards, they pour their contents into a reservoir . . .', which suggests that they were tilted at such an angle that their contents did not begin to pour out until they had passed 'top dead centre'. If this was in fact achieved, it represents a big improvement on the round earthenware pots still to be seen on wheels of this type in some Mediterranean countries.

The design was forgotten, and re-discovered in the Middle Ages, since when it has been known as a 'Persian wheel'. Nowadays these wheels are usually turned by animals. Though Vitruvius makes no mention of this, there are clear indications in papyrus documents that oxen or donkeys were used in Egypt by the Greek-speaking communities in the second century A.D., if not earlier. In one document, the writer excuses himself for not fulfilling an irrigation contract 'because there was not enough fodder for the animals'. In another, arrangements are made for supplying water to 'the men who drive the animals round', and another deals with the supply and purchase of 'connecting straps' *(zeukteriae)* for harnessing the animals. In all these documents the pump is called simply *mēchanē*, which could mean any type but most probably means either a 'drum' or a bucket-wheel. Since each of these requires a horizontal drive, and the animals could only have turned a capstan with vertical axle, some form of gearing must have been fitted, either a bevel gear or—more probably—a primitive crown wheel and pinion. (The only possible alternative might be a treadmill big enough for a donkey to walk around inside. Such an apparatus was used for hauling up water from a deep well at Greys Court, near Henley, until the First World War, and can still be seen there.) In another ancient document there is an account of a fracas in which some parts of a pump were set on fire and partly destroyed. The word used—*ergatai*—might mean the bars of the capstan, or (possibly) the crown wheel and pinion.

The obvious advantage of this type of pump over the drum is that a wheel of the same diameter can raise water twice as high— perhaps even more, but it has one disadvantage which the drum does not share. As the wheel revolves, the buckets are tilting all the time, and there is a danger that some of their contents will be

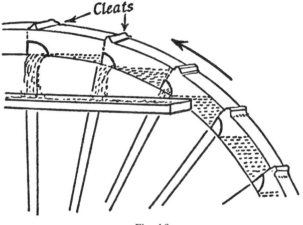

Fig. 16

spilt, either before they reach the launder, or after they have passed it, or both. Vitruvius may have provided one solution by choice of shape for the buckets. Another ingenious one can be seen from the remains of a Roman pump of this type found (during the early 1920's) in the Northern workings of the Rio Tinto mine in Spain. The wheel itself, 14ft 10in (just over 4.5m) in diameter, was made entirely of wood—the hubs and bearing of oak, and the rest of pine—and wooden dowels were used throughout the construction, presumably because iron nails would have rusted too rapidly in the wet conditions. The axle was made of bronze, square in section with its ends rounded where they fitted into the bearings. There were 24 spokes running alternately from each hub to the inside of the rim, which was made hollow and rectangular in cross-section (as shown in Fig. 16), the internal measurements being about 7 × 5in (18 × 13cm). The whole circumference was divided up by partitions into 24 compartments.

The openings through which the water entered each compartment from the sump and left it at the top of the circuit were on the side of the wheel, and roughly quadrant-shaped, their straight edges being on the inner rim and the leading end of the compartment (see Fig. 16). This arrangement ensures minimum spillage before, and complete emptying of the compartments after, 'top dead centre'. The water was collected from the outlets in a launder, which was fixed as close as possible beside the wheel without rubbing on it, and as high as could be managed without serious spillage. Palmer's estimate of 25% loss seems rather pessimistic, but may not be far out. Combined with power losses from friction in the axle bearings and the paddle effect of the rim passing through the water in the sump, this brought the efficiency of his reconstructed model down to just over 60%. Its output was nearly 19 gall (about 93.5l) per minute, assuming a theoretical lift of 12ft (3.65m). The diameter of the wheel was more than this, but when allowance is made for the width of the rim dipping under water, and for the launder being low enough to catch most of the outflow, the actual lift was probably about equal to this theoretical figure. The power required to work this wheel— just over 0.1 h.p.—could have been provided by one man over an 8 hour shift.

One of the most ambitious Roman mine-draining systems, the remains of which were found in another part of the Rio Tinto mines, had a succession of eight pairs of these wheels, and was designed to raise the water altogether about 97ft (29.6m). The arrangement is quite ingenious. There is a natural tendency for the wheel to propel the water along the launder, and the flow is in any case pulsating rather than steady. If two wheels were discharging into parallel launders, there would be a tendency for turbulence to be set up in the region where the launders joined, which could only be corrected by increasing the downward slope of the conduit leading to the sump for the next higher pair of wheels, and this in turn would involve losing a few inches of lift. So the two wheels (side by side) are made to revolve in opposite directions, and a single launder runs around both (Fig. 17), so that each wheel propels the water along it in the same direction, and only a very slight fall is necessary. This system would have an impressive output—about 2,400 gallons per hour at a head of nearly 100ft is no mean achievement, but it should be remembered that the total cost must have been considerable, and that it needed the full-time labour of 16 able-bodied men. The ore deposits must have been very rich to justify such an investment.*

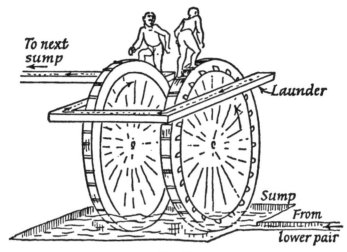

Fig. 17

The second high-lift device is a simple and logical extension of the bucket-wheel, and has the great advantage that the height to which it can raise water it not limited by the diameter of the wheel, but only by the available power, in relation to the quantity required.

*A reconstructed part of one of these wheels can be seen in the 'Roman Life' room at the British Museum. A remarkable feature is that they were apparently made up from 'construction kits', the wood for some of the parts not being native to that part of Spain, but imported from elsewhere. For ease of construction (in the place where they were to be used) some of the side pieces for the rim-buckets were numbered in Roman numerals. They have been wrongly reconstructed in the British Museum. It is quite clear that the pieces would be consistently numbered either on the outside or on the inside, and as a result of ignoring this, the buckets have been reconstructed with outlet holes on both sides of the wheel. The rate of outflow through one side would have been ample, and a second outlet would have required a second launder, and would have doubled the wastage by spilling. The method of constructing the hub and of fixing the spokes in it, can be seen quite clearly. The same error is repeated in the drawing of the Dolau Cothi wheel—see G. C. Boon and C. Williams in JRS LVI (1966), 122–7.

Vitruvius gives a very brief sketch of it at the end of the same chapter (X, 4) without giving it a name, but from the passing reference in Hero (p. 207) we gather that its Greek name was *halysis* ('chain'). It had a treadmill on a horizontal axle, and two parallel endless chains were suspended from the axle, with buckets fixed to them

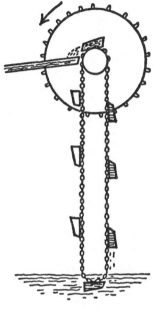

Fig. 18

at intervals. According to Vitruvius, the standard size of bucket was one *congius* (about $5\frac{3}{4}$ pints or $3.3l$). The chains were long enough to hang down to the water level below, so that the buckets dipped in as they reached their lowest point. They were then held upright all the way up to the axle, tipping over automatically as they reached it, and emptying their contents into a conduit (Fig. 18). This arrangement can be made to work more efficiently than on the bucket-wheel, since the buckets do not begin to tip until they actually reach the axle, and then they turn over rapidly and decisively. Waste by spillage can be reduced to a negligible minimum.

There is, however, one serious mechanical problem which must have arisen on this machine. Unless the weight of water being lifted at any one time (i.e. the contents of half the buckets) was small in relation to the total weight of the chains and buckets, which in turn means that the output was small, there must have been a tendency for the chains to slip around the axle. At first sight it might appear that Vitruvius was unaware of this problem, and has neglected to describe some sort of sprocket arrangement on the axle, to engage with alternate links of the chains. This need not have been anything more sophisticated than a set of headless nails, driven into the axle at the appropriate spacing. As an additional measure to prevent slippage, the shaft may have had longitudinal grooves, suitably spaced so that when the horizontal links of the chains engaged with the sprockets, the vertical ones embedded themselves in the grooves. It should be remembered that Vitruvius' description is very brief indeed, and he is concerned only with the structure as a whole, not with details of this kind. He certainly does seem to be talking about a pair of iron link-chains *(duplex ferrea catena)*; these were quite common in Roman times, and a number of examples have been found on Romano-British sites — one particularly interesting example, which happens to be a chain for a well-bucket, can be seen in the Priest's House Museum in Wimborne, Dorset. They were made by blacksmiths using the same method, known as 'shutting', that survived well into the 20th century. It involves heating a short iron rod to dull red heat, hammering it into a ring with its ends overlapping slightly, then heating them to bright red heat and hammering them together to form a weld. One important consideration is that the weight of the chains was not a problem in the operation of the water-raising machine, since the descending halves would balance out the ascending ones, and the only load to be lifted would be the water in the ascending buckets. In my hypothetical machine described on p. 74 this would be about 85 lbs (39 kg), and there would be no problem (apart from the expense) in making the chains strong enough to cope with this order of load without stretching. The more serious problem would be friction between the links as they passed around the shaft, and the wear which it would cause on the small areas of the links which were in contact.

The problem of slippage on the shaft was a severe one, which had to be solved completely. Even if it was very slight, the two chains might not slip to the same extent, which would cause the

buckets to hang askew and jam on the shaft. Philo's account of
the repeater catapult (see p.123–6), may perhaps offer a solution
to the problem. The shaft may have been faceted—in the form of
a pentagon or hexagon, and the chain made with straps (not links)
to fit the facets. The buckets could have been suspended from
horizontal rods fixed between corresponding straps (see Fig. 19).
This would eliminate slipping altogether, and ensure that the buckets hung level. It would also eliminate the problem of power-drive
on the lower shaft.

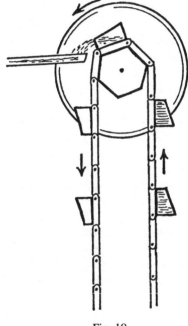

Fig. 19

The bucket-chain must almost certainly have been more expensive to construct than the bucket-wheel (which could be made
without ironwork), and only in situations where the wheel could
not be used would the chain be preferred. Two such situations
would be (a) when the total lift had to be more than about 20ft
(6m)—it would be difficult to make a wheel bigger than that in
diameter—or (b) where the source of water was a well or other

place of restricted access, where it would be impossible to dig a pit big enough for a wheel.

Let us try to imagine a bucket-chain in action—once again, by conjecture based on a number of arbitrary assumptions. Suppose the chains to be just over 110ft (33.6m) long, 'wrapped around' a shaft 15¾in (40cm) in diameter. The maximum possible lift, with the buckets dipping just below the water, would be about 52ft (slightly less than 16m). On chains of this length, twelve buckets could be attached at intervals of just over 9ft (2.8m); if the buckets were of the size recommended by Vitruvius (one *congius*), one complete revolution of the chains would raise just over 8½ gall (39.6l) of water. To accomplish this, the axle would have to be rotated about 27 times which, unless the treadmill was quite small, might well take as long as three minutes. At that speed the output of just under 3 gall (13.2l) per minute would represent one man's work at an efficiency of just under 50%. The implications of these dramatic, but by no means fanciful figures, are clear. It would be impossible to irrigate more than a tiny area of cultivated land by this means, and it is most unlikely that the produce of such an area would balance, or come anywhere near to balancing, the cost of installing and running the machinery. On the other hand, it might supply the domestic needs of a modest villa—drinking, cooking, washing and sanitation—even if these had to be met from a well rather deeper than usual. If worked for two hours a day it would provide 350 gall (nearly 1600l).

There is a means of escape from these severe limitations where the source of water is a river or fast-flowing stream, which can provide an alternative to man-power. At the start of the next chapter (X, 5) Vitruvius describes how a water-wheel was used to work a bucket-chain. It was apparently a paddle-wheel of the under-shot type, which, if efficiently constructed might well deliver more power than one man. A few extra problems might arise; the power was applied at the bottom of the chain circuit instead of the top, and whereas the weight of the chains and buckets provides extra grip on the top axle, it would not help here. In fact, it is almost certain that the faceted shaft and strap-chain were used. But the great advantage of this system would be that it could be kept running 24 hours a day, without employing any able-bodied men. An old or infirm slave could keep an eye on it, and raise the alarm if anything went wrong. Assuming that the water-wheel developed 'three man-power' (0.3h.p.) and ran continuously, it could perhaps deliver 12,600 gall (54,000*l*) per day.

In what situations would it be desirable to raise a comparatively small amount of water from a running stream to a high head? There are two which spring to mind. One is a villa built on rising ground near a river, with domestic supplies pumped up by one of these machines, and perhaps reservoirs *(stagna)* to be kept filled. The other is where a river flows across a plain, and it is necessary to provide irrigation some distance away from the nearest point on the river. Using the machine described above, water could be raised to the top of a tower on the river bank and taken thence by a sloping conduit (or rather, a miniature aqueduct). With a gradient of 1 in 200 (not uncommon in aqueducts) the water could be conveyed over a distance of nearly 2 miles (about 3.2km) to the land requiring irrigation.

Finally, there was a fifth type of pump for which we have good literary evidence and some archaeological remains. This is the force-pump, with pistons, cylinders and valves, called in antiquity a 'Ctesibian machine', after its inventor, Ctesibius (third century B.C.). A description of this type of pump is given by Vitruvius (X, 7) and by Hero of Alexandria, whose fire-engine made use of it (*Pneumatica* I, 28). Both these writers also give descriptions of an organ, in which a piston-and-cylinder pump was used to force air into the reservoir (*Pneumatica* I. 42, Vitruvius X, 8) The *pneumatikon organon* referred to by Pliny (*Nat. Hist.* 19, 20), for watering vegetable gardens from a well, was almost certainly a pump of this type.

The design was as follows. There were two vertical cylinders, their pistons worked reciprocally by a rocker-arm (Fig. 20). If this was accessible at ground level (as in the fire engine), a handle was fitted to an extension of the rocker-arm, at one end or both ends. If the pump was submerged, or down a well, a wooden push-rod must have been attached to one end of the rocker-arm. The connecting rods pivoted on the rocker-arm at their 'big ends', and it is interesting to note that Hero says (*à propos* the organ pump, at the very end of Chapter 42) that it is better to have a pivot at the piston end of the rod, to avoid lateral motion of the piston. This pivot corresponds exactly to the little end and gudgeon pin in a modern petrol engine, and would help greatly to reduce wear on the piston and cylinder, especially when the connecting rods were short. In the absence of any flexible tube or connection, the cylinders must have been rigidly mounted.

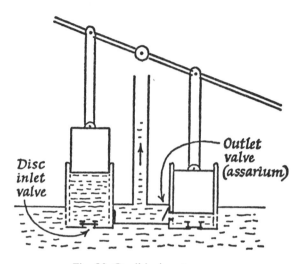

Fig. 20. Ctesibius's water pump.

Both Hero and Vitruvius specify that the pistons and cylinders should be made of bronze, this being probably the strongest metal which could be worked to the necessary degree of precision with the tools then available. It would be very nice to know more details of the manufacturing process. Hero merely says that the cylinders 'have their inner surfaces worked to (fit) the pistons, so that there is no by-pass', and that this was done on a lathe. Vitruvius uses the phrase *oleo subacti,* which is difficult to interpret. *Subactus* is used elsewhere to mean precision-turned on a lathe. If that is the meaning here, then *oleo* must mean that olive oil was used to lubricate the cutting tool. But the more natural meaning of *subactus* would be 'forced', which might mean that the final trimming was done by ramming the piston into the cylinder, using olive oil as a lubricant so that each was 'trued up' by the other. The technique is known as 'lapping'. All this, however, is very speculative, and it should be added that olive oil is a poor lubricant for metal surfaces and would do little to ease the movement of the pistons when the pump was working. In Vitruvius' description of the organ, where the pump is used to compress air, it is stated that the pistons should be 'wrapped in sheep's hide with the wool still on it', but there is no mention in Hero or Vitruvius of a leather washer on the piston of the water pump, which is surprising.

The inlet ducts were in the form of round holes in the bases of the cylinders, over each of which (inside the cylinder) was a non-return valve of a very simple type. It consisted of a disc, which, when held against the inlet during the compression stroke, more or less sealed it off. The disc was held loosely in position by four pins, which passed through holes near its edge, so that it could release the seal during intake, but could not become displaced, and would close up immediately under pressure. Hero describes another type of valve which, he says, the Romans called *assarium* ('penny-valve' —an odd choice of name, as it would belong much more appropriately to the disc type. Was Hero's Latin a bit shaky?) This consisted of two bronze plates, about ¾in (2cm) square, one of them having a round hole in its centre. This plate was sealed onto the outlet duct, and the other plate was attached to the first by a hinge running along the top edge (Fig. 21). The two adjacent

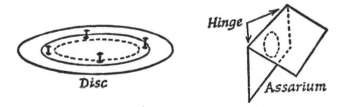

Fig. 21. Pump valves.

surfaces were carefully smoothed to make a good seal. Both this type of valve and the disc type were gravity loaded, and could only be mounted in one position—the disc horizontal and the *assarium* vertical. Reliable spring loading would be very difficult to achieve with the materials then available. The best that Hero could find for the organ key-mechanism was a strip of animal horn.

The outlet pipes come from the inward-facing sides of the cylinders, near the base. Hero's design has a T-piece with a vertical

pipe coming up from the middle, with *assarium* valves on the out-
sides of the cylinders, but because Vitruvius uses outlet valves of
the disc type, it is necessary for each of the outlet pipes in his plan
to curve upwards and enter, vertically, the base of a shallow bowl
(catinum). The valves are fixed on the ends of the pipes, and an
inverted funnel is then placed over the bowl, firmly fixed to it,
with the outlet pipe rising upwards from its spout (Fig. 22).

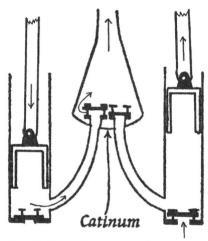

Fig. 22. Vitruvius' outlet valve system.

For water to be drawn in, it was clearly necessary for the bases of
the cylinders to be kept under water. In the fire-engine this would
be done, as it was with the wartime stirrup pump, by tipping buckets of water into the tank (see below), but the whole pump can, if
necessary, be submerged to any reasonable depth. In a well, for
instance, in which the water level rises and falls seasonally, the
pump could be fixed below the lowest normal level, and would
work efficiently whatever the depth at any given time. The only
variation would be in the degree of lift required, and hence in the
power needed to work the pump.

Parts of several pumps of this type have been found, one at Trier
in Germany, and the remains of another, now in the British
Museum, comprise most of the two cylinders, the pistons, the
T-piece and parts of the disc-valves. Parts of another much larger
pump were found in Silchester and are now to be seen in the

Reading Museum.* The whole structure was enclosed in a wooden block (probably a tree-stump), most of which is preserved—due, no doubt, to the fact that it was under water—and the carved-out portions where the cylinders, outlet pipes and valve chamber were fixed can be clearly seen. The closeness with which they fit Vitruvius' description is quite remarkable. Two long lead tubes also survive, one still in position in the block and another apparently identical, with traces of solder on it. These were almost certainly the cylinders. The pistons are not preserved, but whatever their design they could have travelled upwards to within about 6 in (15 cm) of the tops of the cylinders. The inlet valves, mounted on wooden plugs in the bases of the cylinders, would allow a stroke of about 12 in (30.5 cm). The cylinder bore was about 3 in (7.6 cm), so the capacity of each cylinder would have been about 1.4 l., and a double stroke would theoretically have delivered 2.8 l. So if the pump were operated at 15 double strokes per minute (which is fairly slow) it would in theory have delivered 42 l (9$\frac{1}{4}$ gallons) per minute. However, the efficiency must have been well below 100%, due to friction in the cylinders (the pistons may have had leather washers), and leakage in the valves. A reasonable guess would suggest about 50–60% efficiency, and an output of about 21–25 l (4$\frac{1}{2}$–5$\frac{1}{2}$ gallons) per minute. To deliver this at a head of 16 ft (4.9 m), which is about the average for the wells in Silchester, would require a power input of a mere 1/20th horsepower, which one man could provide without much effort.

Hero's fire-engine, described in Book I of his *Pneumatica*, Chapter 28, involves a two-cylinder force-pump. It has a water tank, for which Hero gives no measurements, neither does he actually say that it was mounted on wheels, though it almost certainly was. Presumably he thought that the outward appearance was familiar to his readers, and concerned himself only with the unseen working parts. The cylinders of the force-pump were fixed to the base of the tank, and the pistons were worked by a rocker arm pivoted on a post in the centre. The outlet pipe rose vertically from the T-piece, and was fitted with a device to enable the nozzle to be tilted up or down, and swivelled round in any direction—obviously a vital feature, in the absence of flexible hoses. The

*See Frederick Davies, appendix in *Archaeologia* 55 (Part 1) 254–6. G. C. Boon, *Roman Silchester* (Max Parrish, London (1957) 159–61) and Trierei Zeitschrift XXV, 109–21.

device has two joints of the kind which Hero calls 'sleeved' *(synesmerismenon)*—i.e. with one pipe fitted inside another, tightly enough to prevent leakage, but loosely enough to allow the unfixed pipe to twist around. The elevation adjustment is made by the arrangement shown in Fig. 23, where the joints at A and B are 'sleeved'. Hero then explains that the nozzle must be able to turn horizontally, 'otherwise', he says, 'the whole apparatus will have to be turned round, which is slow and pretty useless in an

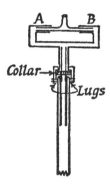

Fig. 23. Hero's nozzle system.

emergency'. So the vertical outlet pipe from the pump is also 'sleeved', enabling the whole top part of the nozzle assembly to turn round through 360°. There is still one further refinement. A collar is fixed around the outer pipe, and L-shaped lugs are fixed on the inner one to engage with it (Fig. 23). 'This', says Hero, 'prevents the upper pipe from being blown right off the apparatus by the water pressure'. In view of these eminently practical comments (what could sound more like bitter experience than the second?) the suggestion that the whole machine was an unusable armchair invention seems rather absurd; though it becomes more understandable, if not excusable, when one looks at the ridiculous diagrams of it in a number of textbooks.

Remains of a force-pump with vertical outlet pipe and rotating nozzle, which closely fits Hero's description, were found in 1889 in the Sotiel Coronado mine in Spain, some distance below ground. In archaeological textbooks and articles the device is usually referred to as the 'Valverde Huelva pump', and it is housed in the National Archaeological Museum in Madrid. The nozzle can be turned horizontally and vertically, and the only difference between this pump and Hero's design—a trivial one—lies in the outlet valves. Hero uses the *assarium* type, with a vertical hinged flap, while the Valverde Huelva pump has both outlet and inlet valves mounted horizontally. They are of the disc type, and a little more sophisticated than Hero's disc valves, in that each is enclosed in a small cylinder. Each disc has a short vertical stem which rises vertically from its centre, and the top of the cylinder has a central

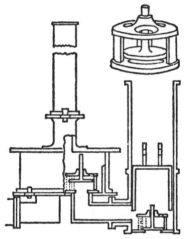

Fig. 24. Valverde Huelva pump.

hole through which the stem slides up and down—a valve-guide, in fact. Around this central hole are other larger ones through which the water passes when the valve is open (Fig. 24). In order to make the mounting of the outlet valve horizontal, the outlet pipe from each cylinder has a right-angle bend, and leads into a chamber which corresponds to the *catinum* in Vitruvius' design.

More recently, four bronze pumps of a slightly different design were found in the 'Dramont D' wreck—a Roman merchant ship dating from about the middle of the first century A.D.* It is not clear whether these pumps were part of the merchandize being carried, or whether they were the ship's bilge-pumps. If so, they were apparently not in use when the ship went down. They show a

*See G. Rouanet, *Etude de quatre pompes à eau Romaines,* in Cahiers d'Archéologie Subaquatique III (1974) 49–79.

curious feature, which recalls the bucket-wheels in Spanish mines. The various parts of the four pumps have distinguishing marks (dots, arrows, etc.) which appear to be guide-marks for the correct assembly of the parts. These marks are all the more important because the pistons and cylinders are not of exactly standard bore, and therefore are not interchangeable. At what stage they were put on, and for whose benefit, is a matter for argument. Perhaps they were put on in the actual foundry where the bronze casting was done, and used by the craftsmen who assembled the parts (mainly by soldering), even though they were probably in another part of the same factory.

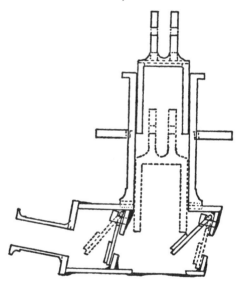

Fig. 25. Pump found on Roman merchant ship.

These pumps have yet another different arrangement for the valves. Each cylinder is mounted on a rectangular box, with the inlet valve at one end, the outlet pipe at the other, and the outlet valve on a partition between the cylinder wall and the outlet end (Fig. 25). Both ends, and the partition, slope at an angle to the vertical of about 15°–20°. The valves are of the *assarium* type, with a flap which hangs from two hook-shaped brackets at the top. As each valve is on the upward-facing side of the sloping 'bed', it naturally tends to fall shut unless forced open by the flow of water.

Accurate measurements of these pumps give a new insight into the competence of ancient metal-workers, and bring sharply into question some of the traditional arguments to the effect that inadequate technology limited the progress of applied science. The measurements of piston and cylinder in each pump are as follows:

		Piston outside diam (mm)	Cylinder inside diam (mm)	Clearance all-round (mm)
Pump	No. 1	43.5	44	0.25
	No. 2	42.8	43.5	0.35
	No. 3	42.9	43.1	0.1
	No. 4	44	44.3	0.15

The inner surface of each cylinder and the outer of each piston has been highly polished, probably with some form of grinding paste, and if it is coated with heavy grease makes a very good fit. Tests suggest that the overall efficiency of one pump, including the valves, was about 95%. Its output, pumping at a rate of one stroke (up-and-down) per second, was about 140 gall (630*l*) per hour.

Of cruder construction, but on a more impressive scale, is a pumping system discovered at St Malo in Brittany in 1971.* This included an eight-cylinder force-pump. The 'cylinder block' was made from an oak beam, the cylinders themselves being drilled out in a line along it. In the language of automobile-makers, it is a 'straight eight'. No trace of the pistons survives, and of the valve mechanism we have only a set of eight short wooden tubes, which fitted into ducts at the bases of the cylinders. Nor is there any trace of the mechanism for working the pistons, and it is impossible to tell whether they all rose and fell together, or four-and-four alternately. Even the purpose of this pump is obscure. It was found in a square tank connected to a small square reservoir, down near the sea shore (in fact, the site is now below sea level). Perhaps by some quirk of nature there was a supply of fresh water there, or perhaps the pump served to fill up salt-pans. We shall probably never know.

*See R. Sanquer in *Gallia*, Tome 31 (1973) Fasc. 2, pp. 355–60.

4

Cranes and Hoists

WE have various sources of information on ancient cranes. Actual remains of the machines are almost completely lacking, but there is abundant evidence in the form of buildings erected by means of cranes, some of which must have had a quite remarkable lifting capacity. The architrave sections of the Parthenon, for example, weigh something in the region of 9 tons each, and had to be raised to a height of about 34ft (10.5m) for positioning on the columns. Most of the columns are built up of 11 drums, each weighing about 8 tons, which had to be lowered accurately onto a central spigot. This example, from the latter half of the fifth century B.C., is chosen as the best known, though it is neither the oldest nor the biggest. Apart from such evidence, we have an extended Latin account of two types of crane (Vitruvius X, 2, 1–10) and a detailed relief sculpture, probably of the early second century A.D. It occupied a panel in the family tomb of the Haterii, one of whom was apparently a building contractor.* This sculpture has a number of interesting features. The carving was done by a mason who was familiar with the crane itself and had an eye for detail. For instance, two men are shown tying a rope around the top of the jib, and the carving is so accurate that it is possible to identify the knot as a reef-knot. As an artist, he leaves much to be desired. His perspective is hopelessly bad, and the buildings are lumped together in an almost surrealistic way. But this does not detract from the value of his evidence on technical matters.

Vitruvius' description of the first type of crane is as follows. Two beams are required for the jib, their thickness depending on the maximum probable load. They are fixed together at the top with an iron bracket, and separated at the base, like an inverted V

*Illustrated in many places, e.g. G. E. R. Lloyd *Greek Science after Aristotle* (in this series) Chatto & Windus, London 1973, fig. 24, p. 110. The best photographs are in Atti della Accademia Nazionale dei Lincei, series VIII vol 13 (1968), figs. 15 and 17.

(hence the modern name 'shear-legs'). Ropes are attached to the head of this jib, and arranged 'all round' to keep it steady. A pulley block is suspended from the top with two wheels, one above the other; the hoisting rope passes over the higher of these, down and around the (single) wheel of the lower block, which is attached to the load, up again and around the lower pulley of the upper block, and down again to an eye on the lower block (Fig. 26).

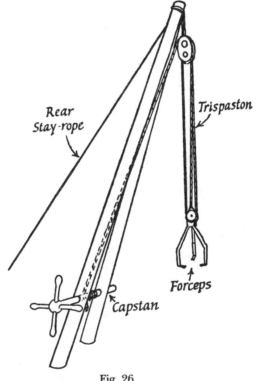

Fig. 26

The other end is brought down between the legs of the jib to a windlass turned by handspikes (four, all at right-angles). From the lower pulley block iron forceps are suspended, the teeth of which fit into holes in the blocks of stone to be lifted. This arrangement is called a 'triple-haul' (in Greek, *trispaston*). If the upper block has three pulleys and the lower block two, it is called a 'quintuple-haul' (*pentaspaston*).

If the crane is required to lift bigger loads, the beams for the jib must be longer and thicker, and the other components proportionately strengthened. Vitruvius then describes a technique for raising the jib from the ground, using a second windlass with a stay-rope passing over a pulley at the top of the jib, over the 'shoulder-blades' (*scapulae,* presumably raised projections on the

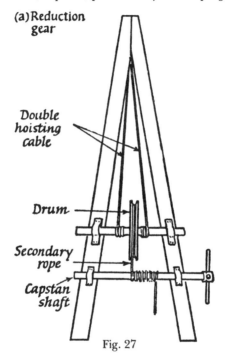

(a) Reduction gear

Double hoisting cable

Drum

Secondary rope

Capstan shaft

Fig. 27

back of the jib, to reduce the initial mechanical disadvantage), and via a double-pulley system to an anchor-point some distance away or, if there is none available, to a set of inclined piles driven into the ground. Once the jib is raised, adjustable stay-ropes on all sides can be attached, each with its own anchor-point. If the windlass and quintuple-pulley system are not powerful enough to raise the load, an additional reduction gear is used. In place of the windlass there is a shaft passing through a drum, fixed at its centre. The hoisting cable is doubled, its two ends winding onto the shaft to the right and left of the drum, and the pulley blocks have pairs of wheels side by side instead of single wheels. A secondary rope is wound round the circumference of the drum, and taken to a wind-lass, probably mounted below on the jib (Fig. 27). Alternatively, the drum can be made large enough to act as a treadmill, worked by men from inside. If so, it would be mounted at one side of the jib, on the end of its shaft (Fig. 28).

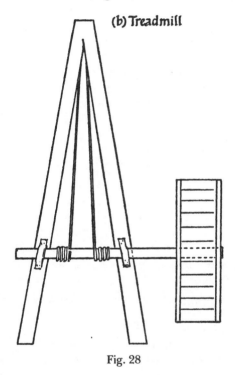

(b) Treadmill

Fig. 28

This probably represents a fairly advanced stage in the development of crane design. The later centuries of the Roman Empire may have improved on the details, but the fact that an illustration of a century later fits this description almost exactly suggests that the design was standardized. It had certain merits and limitations.

The forked jib, unless some pivot mechanism was fitted on the base (and Vitruvius does not mention any), could only swing forward and backward along a straight line over a limited distance. If the jib is raised too near the vertical, the load strikes against it, and may damage it or the lifting tackle. As the jib is lowered forward from the vertical, the stress on the rear stay-ropes increases, until at an angle of about 30° from the vertical (depending on the distance of the anchor-point) it becomes equal to the load being raised, and beyond that point it exceeds the load. So the forward tilt of the jib, and hence the range over which the crane can move its load, are limited by the strength of the stay-ropes and of the mechanism for adjusting them. This is why Vitruvius says that the jib must be longer for a greater load, since it can then move its load forward over the same distance with less deviation from the vertical.

The second design described by Vitruvius has a single-beam jib, and he points out that it can be swung sideways as well as forwards. However, its handling capacity was probably a good deal less. At the opposite extreme, a tripod derrick, mentioned by Hero in the *Mechanica* (III, 4) would have no mobility at all.

The stay-ropes at each side are required to prevent toppling, especially when the load is not initially central and swings about when lifted from the ground, but it is absolutely essential to have another running forward from the jib-head, as is shown in the relief illustration. The danger is that as the jib is tilted upwards towards the vertical, the load on the rear stay-ropes diminishes quite suddenly to zero. Without the forward stay-ropes there is nothing to prevent an over-enthusiastic heave from bringing the jib and the load right over backwards and down onto the crew.

The disadvantages of the multiple-pulley system lie in the loss of energy through friction in the pulleys and creep in the rope. The advantages are that it not only provides a reduction ratio of 5:1 (with a *pentaspaston),* but also distributes the weight of the load evenly between five sections of the lifting cable, so that a rope with a capacity of (say) $\frac{1}{2}$ ton can be used to lift up to $2\frac{1}{2}$ tons, or 5 tons if doubled. The gearbox system described by Hero of Alexandria* has a reduction ratio of 200:1—or more, if a worm gear is used —and could be used instead, but the load is attached directly to the final shaft, and a single thickness of extremely strong rope would have to bear the entire load. It is difficult to support the view of some scholars that this gearbox was merely a pipe-dream —pipe-dreamers do not allow for friction in the gears, as Hero does. However, it may well be that for practical reasons the older and less sophisticated pulley system was preferred. The rope-driven reduction gear is simple, and was presumably as

*see pp. 206–7.

cheap to make as a gear system, but it had its problems. To raise the load 10ft (3m) on a five-pulley system, 50ft (15m) of the hoisting cable must be wound up, and supposing the windlass shaft to be about 6in (15cm) in diameter, this would require about 32 revolutions of the shaft, ending up with a lot of rope wrapped around it. If the drum were 3ft (90cm) in diameter giving a reduction ratio of 6:1, 300ft (90m) of the secondary rope would have to be unwound from it—32 turns again, though of thinner rope in this case. If the second windlass were the same size as the shaft, it would have to be turned nearly 200 times, and the bulk of rope around it would become very large. Perhaps the rope was wrapped a few times around the windlass shaft, held under tension to avoid slipping, and stowed in a heap instead of being wound onto the shaft.

If the handspikes on the final windlass were 3ft (1m) long the total reduction ratio would be 180:1. Even allowing for considerable loss of energy in the pulleys and the rope—say 30%—one able-bodied workman could probably hoist at least 2 tons with it. It would take some time, perhaps half an hour to raise it 10ft, but this could be speeded up by extra men, up to four if there were handspikes at each end of the windlass.

It seems highly probable that the weak point of the whole apparatus was the attachment between the lower pulley block and the load. The 'iron forceps' *(ferrei forfices)* are not visible in the illustration. But we have a brief reference to them in Hero's *Mechanica* III, 7. The details are by no means clear, but apparently the device was called a 'crab', and had three or four 'jaws' (three for column drums, and four for rectangular stones) with bent ends. These were either pushed under the bottom edge of the stone to be lifted, or into holes specially cut in the sides. Then wooden cross-pieces were fixed on (perhaps tied on with ropes?) to prevent the jaws coming apart and allowing the stone to slip through. The holes in the stone blocks must have been difficult to conceal unless the stone was deeply carved after being laid in place, which seems unlikely. If they were cut in the ends of the block, they could be concealed by adjoining blocks in ashlar masonry, but they would have to be laid to one side of the final position for the jaws to be removed, and then shifted along by other means. Rollers were apparently used for this on some occasions. Moreover, the danger of a block splitting in mid-air and falling from the forceps must have been ever-present. Hero points this out as one of the hazards for the crane crew.

Some other methods of attaching the load are known from ancient evidence. One was to leave projecting lugs on the sides of stone blocks or column drums, which could be trimmed off when the stone was laid in place. On some column drums the lugs appear large enough for ropes to be tied around. The fact that they are below the centre of gravity does not constitute a problem. Once the hoisting ropes were looped over the lugs, two other ropes could be tied around the circumference of the drum, one just above the lugs to keep the hoisting-ropes from slipping off them, and the other near the top of the drum to prevent it from tipping over when lifted. In fact, it would be safer to lift it in this way than to have the lugs higher up, since the weight of the portion of the drum below the lugs would put an extension stress on the stone, and might crack it if there were any flaws in the centre. Alternatively, the lugs might have provided a grip for forceps.

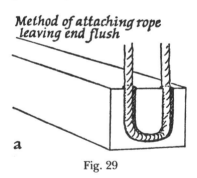

Fig. 29

Another method of attaching the load was to make a sling of ropes passing under the stone. The problem here, of course, is to get the sling away when the stone is laid in place. When working with small blocks as part of an ashlar wall, crowbars could be used to jack up one end while the ropes were untied, and marks (known as pry-holes) have been found on a number of blocks which suggest that this has been done. But two other methods were devised in antiquity for dealing with the problem. One was to cut a deep U-shaped groove in each end of a block (Fig. 29). The ropes passed through these grooves and were prevented from slipping out by a slight flange on the upper wall of the groove.

The blocks could be laid flush together, and the ropes untied and removed. This ingenious method has two limitations. As Greek ashlar stonework was not normally cemented, it was possible for water to gather in the groove which, in exceptionally cold conditions, might freeze, expand and so displace the stones. Secondly, the method could only be used on adjacent surfaces which were eventually to be concealed, and was not suitable for corner-pieces, the ends of courses, or (as a rule) exposed top courses.

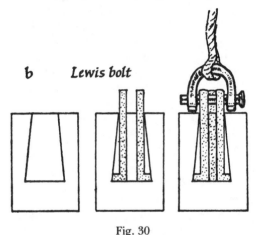

Fig. 30

Yet another method was to use what is known as a 'lewis bolt', or a 'lewis iron', or just a 'lewis'. We have no actual remains of these from ancient times, but a brief reference in the next chapter of Hero's *Mechanica* (III, 8), taken with the archaeological evidence, makes it clear that they were essentially the same as modern ones. A rectangular hole has to be cut into the top surface of the block to be lifted, of the order of 1ft long by 5in wide (30cm × 12cm). As it goes down through the stone, it gets wider and wider (Fig. 30), its sides sloping out at an angle of about 4–5°, in a dovetail shape. The bolt itself is in the form of three iron bars, two of them bent over at their bottom ends 'in the form of the Greek letter Γ', as Hero puts it. These two were inserted into the hole, the bent ends facing outwards, and the third (flat) bar was then pushed between them, holding them apart and preventing them from slipping out of the socket. To hold them all together as one together as one piece, holes were drilled through their top ends and a pin pushed through. The most likely method of attaching the hoisting cable was by a stirrup with holes in its ends, through which this pin also passed (Fig. 30). Hero notes that the quality of iron used for the bars must be carefully checked. If it is too hard (forged for too long with a heavy hammer?) it will crack, and if too soft (annealed?) it will bend. In either event, the failure is most likely to occur at the bends in the outer bars. A happy medium between too hard and too soft must be found.

The advantages of the lewis iron are obvious. It can be used to place a block on a flat bed, without putting anything underneath it or having to move it into its final position, but its disadvantages are serious. Cutting the holes is a long and tedious business, requiring some skill. When under load, the lewis iron acts as a wedge, and the risk of splitting the block (already weakened by the cutting of the sockets) is considerable. This is why at least two bolts were normally used in all but the smallest blocks, to distribute the stress. Finally, when the job is done, the block is left with a large unsightly hole, which has to be covered over or filled in by some means. In view of all this, it is not surprising that evidence for the use of the lewis iron is not copious, and is almost entirely confined to buildings which have been dismantled and rebuilt after damage (e.g. by earthquake). In such situations there is no alternative method, since the lugs have been removed.

The heaviest blocks used in any building are usually the architraves and lintels, which can be hoisted into place using a rope sling without raising a problem over its removal. They do, however, require very accurate placing, and this may be difficult if the weight is very great. An ingenious method was devised by Chersiphron, the architect of the temple of Artemis (Diana) at Ephesus in the sixth century B.C. He had a mound of sandbags erected between the pillars, rising a little above the capitals, and had the architrave 'raised onto it'. Some scholars have taken this to mean that the sandbags formed an inclined plane, and the architrave was hauled up it. It seems rather incredible, however, that a pile of sandbags, its width limited to the space between columns, could have been stable enough to stand this. Pliny's text at *Nat. Hist.* 36, 97ff. (the source of our information) is doubtful, and could mean 'a gentle slope' or 'a soft support'. It is much more likely that the architrave was hoisted by crane and placed roughly in position, when it could then be adjusted very finely by letting sand out of the lower bags, thus causing the pile to settle very gradually in the required direction.

The design of the crane illustrated on the monument of the Haterii is almost exactly in accordance with Vitruvius' description. The crane has a quintuple pulley system on the hoisting cable, and forward and rear stay-ropes, each with a triple pulley. The pulleys are each constructed from two iron plates, held together by the pulley axles and, in the main hoisting pulley, by a tie-bar at the top, around which the strop passes. The very scanty remains of ancient pulleys which do survive are made of wood.* The machine is not actually in use, but being prepared for use, and the main task of the two men 'up aloft' would normally be to over-haul the pulleys. They are shown tying a rope around what looks like a wicker basket placed upside-down over the tip of the jib, and there is a clump of foliage behind, but it is impossible to say whether it is being tied onto the crane (in some kind of 'topping-out' ceremony?) or whether it is merely for decoration in the corner of the sculpture. The basket may have been a weather-protection for the wooden jib or, more probably, a buffer to avoid damage to stone carving in case the jib accidentally struck a completed part of the building. The workmen are barefoot—a sensible precaution for climbing around on ropes and pulleys—and one of them appears to be wearing a leather cap—a forerunner, perhaps, of the crash helmet.

Five men are shown climbing about on the wheel. It would be reasonable to suppose that in actual use two men would provide the main lifting power (perhaps three for an extra heavy load), and one would stay at or near the bottom, in communication with a foreman up on the wall or building who could see the exact position of the load. The third man below would make the fine adjustments of height by his own weight, in close co-ordination with the other two. To do this properly would require a lot of skill and extended practice—nothing could be easier than to bring about a very serious accident. No matter how plentiful the supply of slaves may have been, a trained crew must always have been at a premium, and very difficult to replace.

The second type of crane described by Vitruvius (for which we have no ancient illustration) had a single-beam jib instead of

*J. W. Shaw, *A Double-Sheaved Pulley Block from Cenchreae,* Hesperia XXXVI/4 (1967) 389–401.

shear-legs. It was held upright, or tilted forwards or sideways as required, by four stay-ropes, and three hoisting cables, each with its own quintuple pulley system, passed over pulleys at the base of the jib and were hauled by three teams of men. By dispensing with the windlass the speed of operation was greatly increased, but it would clearly require, as Vitruvius says, a highly trained and skilled crew to work it successfully.

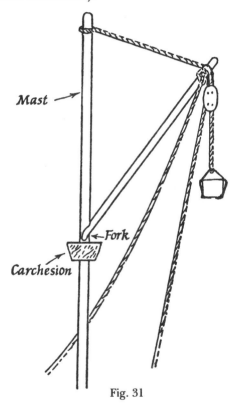

Fig. 31

At the very end of the chapter, Vitruvius casually remarks that other types of crane are used for various purposes, particularly on docks for loading and unloading ships, 'some of them upright (*erectae*) and others horizontal (*planae*), with revolving *carchesia.*' This cryptic sentence requires some examination.

The word *carchesion* can bear a number of meanings. In the context of ship's rigging it seems to mean something like a crow's

nest, or the block in which (or to which) the pulleys for the halyards were fixed. From its shape it also gave its name to a particular kind of drinking-cup. At first sight, therefore, one might be
tempted to assume that Vitruvius is talking about a gaff, such as
can be seen to this day loading and unloading barges and small
vessels. This consists of a spar, with a fork at its base, which rests on
a wooden collar high up on the ship's mast, with the tip of the
spar supported by a rope running to it from the masthead. The
hoisting cable passes over a pulley suspended from the tip of the
spar, and another rope or ropes are used to swing the spar round
sideways, over the ship's hold or over the quay as required (Fig. 31).

There are, however, a number of considerations which cast doubt
on this. The most telling piece of evidence comes from a passage
in the historian Polybius describing the siege of Syracuse by the
Romans in 212 B.C. (He was writing in the second century B.C.,
perhaps 60–70 years after the events.) He describes two ways in
which cranes designed by Archimedes were used as weapons of
war. One was to drop heavy stones or lead weights on approaching ships (the sea came right up to the battlements), and he says
they were 'swung around as required on (or by means of) a
carchesion'. In the absence of any luffing device* (and there is no
evidence for it) it must have been possible to swing the jib forward and backwards, to get the missile directly over the enemy
ship. If it could swing sideways as well, this suggests that *carchesion*
bears the same meaning that it has in relation to a catapult—
namely, a swivel mounting with bearings at each side (Fig. 32).
This is confirmed by the next paragraph, in which Polybius describes a different use of cranes. He tells how landing-parties, on
board ships fitted with arrow-proof protective screens, were prevented from approaching. They were driven back from the prow
of the ship by missiles (stone shot, he says) and then an iron
grappling-hook was lowered on a chain. The man controlling the
jib (literally, 'steering', the word normally used of a ship's helmsman) tried to hook it on the ship's prow. When this was done, he
'drew down' (or perhaps 'pressed down') the opposite end of the

---

*This enables the point of suspension for the load to be moved out
from or in towards the base of the crane. In the modern tower-crane to
be seen on many building sites it takes the form of a small trolley which
runs back and forth on rails along the horizontal boom.

jib, which was inside the wall. (The word used is *pterna* 'heel', used of the butt ends of catapult arms, see p. 117.) This clearly implies that the jib of the crane pivoted, not at its base as the shear-legs did, but (probably) not far from its mid-point. This 'lowering of the butt end', says Polybius, 'lifted the ship's prow out of the water, and stood it up vertically on its stern'. Next, the crane operator 'fastened the machine to make it immovable, and then, by some sort of release mechanism, cast off the grappling-hook and chain. The ships then either capsized, or listed badly, or (Polybius could not resist the temptations of rhetoric) 'became filled with confusion and much sea-water.'

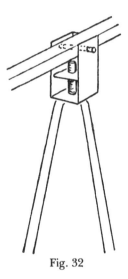

Fig. 32

What are we to make of this account? It could, of course, be complete fiction, but the chances are that it contains a nucleus of truth which has gathered an accretion of fantasy in the telling. In a later account, the Roman commander Marcellus is reported as saying that his ships had been 'sloshed about like wine-ladles'. This, even though it comes from Plutarch (*Marcellus*, 17), might be authentic; Marcellus was trying to save face, and justify his action in abandoning a direct attack and turning instead to a prolonged blockade.

Polybius gives the impression that the ships were hoisted right out of the water, but this is most unlikely and in any case quite unnecessary. If the landing-party had been forced away from the bows, the ship would be down by the stern to begin with, and then it would only be necessary to hoist the bows far enough, and give the hull sufficient list, for water to be shipped over the stern —the laws of hydrostatics (and who should know better than Archimedes?) would do the rest, without any additional pull from the crane. The ship would end up 'standing vertically on its stern', and although most of the crew would by then have jumped or fallen overboard, the buoyancy of the wooden hull would probably find equilibrium with not more than a few feet of the bow out of the water. If the crane had lifted it to a higher level than this, the release of the grappling-hook and chain would cause the hull to plunge quite violently, and even if it righted itself afterwards, the crew would be in the water, the ship awash and useless.

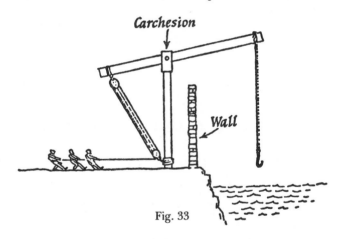

Fig. 33

We are left, then, with an impression of a crane mounted on a tower slightly higher than the fortifications, with a boom on a swivel mounting which could tip up and down, and which, if not pivoted at the centre, might have been balanced with some sort of counter-weight. A type of scaling-ladder *(sambuca)* described by Biton was constructed in precisely this way.* The hoisting arrangements can only be guessed at. One suggestion, which depends heavily on Polybius' expression 'lowered the opposite end, which

*See J. G. Landels, *Ship-Shape and Sambuca-fashion*, J.H.S. LXXXVI (1966), p. 72.

was inside the wall', is that a block-and-tackle was suspended between the far end of the boom and an anchor-point at the base of the tower, and a tug-of-war team pulled on the cable (Fig. 33). Provided the anchor-point was near enough to the axis on which the boom turned, it could be swung around over a limited arc while under load. A similar arrangement (apart from the tug-of-war team!) can be seen on some small mobile cranes used by building contractors. Such a design would fit Vitruvius' description ('horizontal, with revolving *carchesion*'), and would be well suited to the dockside work to which he refers.

5

Catapults

A NUMBER of ancient treatises, and a few surviving illustrations, together present a coherent picture of the designs of various weapons of this type, and of the ways in which those designs were gradually developed and improved. The most important written sources are the works of Hero of Alexandria and Philo of Byzantium.

The precursor of all types of military catapult was the bow, known in the eastern Mediterranean from remote antiquity. Though very simple, the bow is a highly efficient device which can store the energy imparted to it gradually by the action of drawing, and release it very rapidly. Moreover, the geometry of the string is such that the missile is given rapid acceleration and (to borrow a term from other weapons) a high muzzle velocity. The amount of energy which can be stored in a bow is determined by two factors—the stiffness of the bow (i.e. the force required to bend it) and the length of draw. Both of these are in turn subject to human limitations. The first depends on the strength of the bowman's chest, arm and shoulder muscles—a bow requiring a draw force of 100lb (45 kg) is considered to be near the limits of a normal-sized man's capability. It also depends on the ability of the fingers of the right hand to control the string, to hold it during aiming and release it at the right moment. The second factor is limited by the length of the bowman's arms, from the left hand fully extended to the right hand, drawn back beside the right shoulder. It may be said with some truth that the history of military catapults is the history of the engineers' attempts to overcome these limitations.

Their earliest attempt (according to Hero, and there is no reason to think him mistaken) was something like a crossbow, and went by the homely name of 'belly-shooter' (in Greek, *gastraphĕtēs*) for reasons which will become clear. The date of its invention is uncertain, but such evidence as we have suggests that it was not before the fourth century B.C. The weapon was almost certainly not used in the Peloponnesian War between Athens and Sparta (431–404 B.C.), since Thucydides, the contemporary historian of that war, who shows obvious interest in siege-engines and unusual weapons, makes no mention of it.

The basis of this weapon was a bow, probably not much longer, if at all, than an ordinary hand-bow, but rather stiffer—too stiff, in fact, to be drawn by an archer's hand. This brings us right away to a problem which keeps recurring in the study of ancient technology. The designer of the catapult, since he was expounding something new and complicated, described it in detail and provided diagrams, but the bowmaker was a craftsman. He made his bows very much as his father had before him, any improvements being strictly empirical, and he did not write books about it—his apprentices learned their craft by word of mouth and practical demonstration. The engineer is content to specify in his plans a bow of a certain size, without giving any details of precisely how it was made, or even of the materials used. Some social historians interpret this as evidence of class-consciousness— the engineer thinking of himself as an intellectual and of the bowmaker as an artisan. But there are other possible explanations, one of them being that Hero was probably justified in assuming that his readers knew exactly what a bow looked like, and how it was made. A similar problem arises later over the making of sinew-rope.

It is, however, generally agreed that the Greek bows from which catapults were first developed were composite, and there is, at least, some evidence that horn was used in their construction. If so, they were perhaps made in three layers, with a central lath of hard wood, which served as a core, sandwiched between the other two, one of horn on the inside of the arc, which was compressed as the bow was bent, and the other of sinew on the outside, which was stretched. Primitive communities in more recent times have used this type of bow with the sinew either anchored at the ends of the core, or glued onto it. In the absence of any evidence, we can only guess at the Greek methods. The famous passage in Book XXI of the *Odyssey* (388–430) suggests that Odysseus' bow had a layer of horn, and this must refer to the seventh century B.C. at the very latest. Philo also (Chapter 71) speaks of 'horn and some kinds of wood' as the

materials from which bows are made. It is curious that he does
not mention sinew.

The designer of the 'belly-shooter' was faced with three problems; how to draw the powerful bow, how to hold it drawn while
the arrow was aimed, and how to release the bowstring at the right
moment. For an account of his solutions, one might do worse than
quote Hero in a literal translation, with a reconstruction of his
diagram which, unfortunately, has not survived.* (Fig. 34.)

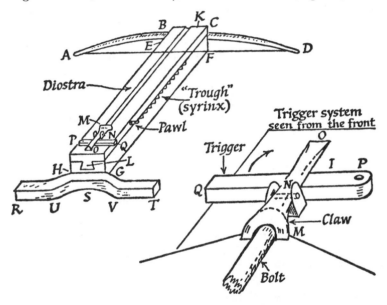

Fig. 34. The bow catapult (*gastraphĕtĕs*, belly shooter).

'Let the bow in question be ABCD, with its curved ends AB and
CD too powerful to be drawn back by human hand. AD is the
bowstring. Fixed to the bow at the centre of its concave curve is a
batten EFGH, which has on its top surface a dovetail groove KL.
Fitted in this groove is a male (dovetail) of the same length, and
fixed to its upper surface is another batten of the same length and

*The passage comes from his *Belopoeica* pp. 75–81 Wescher, Marsden
T.T. pp. 20–22. This translation follows Marsden's text, with a few trivial
omissions and a few explanatory comments. English letters are substituted for the Greek ones on Hero's diagram.

breadth as EFGH; along the centre of its upper surface is a semi-circular groove of the same length as the dovetail KL, in which the missile is placed for firing. On the remaining part of the top of this batten (i.e. between the tail of the missile and the rear end) are two upright iron stanchions, nailed on and joined to each other at their bases, quite close together. Between them is placed an iron claw, bent downwards at the end towards K (the front), and its bent extremity should be divided to form two prongs, as in the so-called *skendylion*. (This was a shipwright's tool, of which little is known; but it was probably something like a tack-lifter or claw-ended jemmy.) The space between the prongs should be wide and deep enough to take the shaft of the missile. A round pin is inserted centrally through the stanchions and the claw. This claw is marked MNO in the diagram, the two prongs at M and the pin I. Below the opposite end NO is placed an iron lever PQ, pivoting on a pin at P, which is fixed vertically on the top surface of the upper batten. When the lever PQ is pushed underneath, it wedges the claw so that it cannot tip up, but when we take hold of the end Q and pull on the lever in the direction of NO, then the front end of the claw—the section MN—can tip upwards. Fixed to the batten EFGH (at its rear end) is a bent bar RSTUV concave at S and convex at U and V.'

Hero then gives a number of standard terms which 'they' (does he mean the designers or the soldiers?) use for the various parts, such as *diōstra* ('slider' or 'push-through') for the upper batten, and *syrinx* ('trough') for the lower. He goes on:

'The machine was constructed as described above. When they wanted to draw the bow, they moved the *diōstra* forwards towards K, until the claw tipped up and rode over the bowstring which, when at rest, was just above the *diōstra*. Then they tipped the claw forwards, and pushed the lever under its rear end, so that it could not tip up again. Next, they propped the front end of the *diōstra*, which was then pushed out at the far end, against a wall or against the ground and, grasping the ends of the bar RSTUV, they pushed their stomachs against the hollow part between U and V and, using the force of their whole bodies, pushed back the *diōstra*, thus pulling back the bowstring and bending the arms of the bow. When they thought the tension was sufficient, they placed the missile in the groove, and released the claw by jerking the lever away from under it, and so the missile was despatched.'

Here we have solutions to two problems—how to draw the powerful bow, by using the whole body instead of only the arms, and how to control and release the bowstring by means of a metal device instead of the archer's fingers. But the third remains, and it may appear to the reader that Hero is guilty of a disastrous omission. In fact, however, he is simply following his usual practice of breaking up the description into clearly defined sections, and avoiding anticipation of what is to come in the next paragraph. He goes on to say that something must be done to prevent the bowstring from pulling the *diōstra* forwards again when the pressure is taken off the stomach-rest. It must be held back until the missile has been placed in the groove and aimed at the given target. This was done by fixing strips (probably of metal, though he does not say so) along the sides of the 'trough' EFGH, with sawtooth ratchets cut in their top edges, and fixing pawls on the sides of the *diōstra,* which rode over the teeth at an oblique angle as the *diōstra* was pushed back. At the appropriate length of draw each of the pawls was propped against the nearest tooth on the ratchet. The fact that the Greek word used for pawl was *korax,* meaning 'raven', might suggest that the pawls were hooked, but this is not necessarily to be assumed.

Hero ends by saying that they called the whole weapon a 'belly-shooter', because the stomach was used to draw back the bowstring. This last remark should apparently be taken to mean that the *gastraphĕtēs* was not so called because it was fired from the stomach. It would in any case have been extremely difficult to aim with any accuracy in that position. The only World War II weapons normally fired from the hip were the Thompson and Sten sub-machine guns, and then only in bursts and at very short range. The *gastraphĕtēs* fired single shots rather slowly and would not be used at close range except in dire emergency. Incidentally, the method of cocking it may remind some old soldiers of painful experiences with the P.I.A.T.

This design did not merely solve the problems of holding and triggering; it also—potentially at least—overcame the restriction on the distance the bowstring could be drawn back, and hence on the length of the bow. Provided it was strong enough not to bend or break under the thrust, the *diōstra* and the groove in which it slid could be made as long as desired. But the problem of bow-stiffness had been only partially solved. Though capable of a stronger thrust than the arms, the stomach—even that of a hefty artilleryman—has its limitations, and the next step was to devise another method of drawing back the *diōstra*, and bending a bigger and better bow. It was quite simply done; instead of being pushed, the *diōstra* was pulled from the rear end by a small windlass and rope. This gave a mechanical advantage equal to the ratio between the radius of the windlass shaft and the length of the handspike. Supposing the shaft to be 4in (10cm) in diameter, and the handspike 20in (50cm) long, the ratio would then be 10:1. If (at a reasonable guess) the artilleryman had been called upon to put a thrust of 200lb (90kg) on the stomach-rest, imagine his relief when, allowing for some friction, a pull of only 22lb (10kg) on the end of the handspike would draw a bow of the same stiffness. Of course, all advantages must be paid for, and the price in this case was speed. Three or four complete turns of a capstan would take longer than a grunt and one hefty shove. Here is a striking example of a situation in which the crank would have been very useful, but Hero calls the handle *skytalē*, which can only mean a rod or spike. Slowness, however, was almost certainly compensated by an advantage which the new design offered at the same time. We must assume that the *gastraphĕtēs* was usually fired over a wall, resting on the top for aiming and pushed against the base for loading. Even if it had some sort of stand or aiming-rest, it would have to be lifted down between each shot. But once the windlass was fitted, it could be kept in the firing position all the time. Incidentally, the weapon was still called *gastraphĕtēs*, though no longer in fact a 'belly-shooter', just as rifle shooting in World War I was still called 'musketry'.

The engineers had now reached a stage at which the best general layout had been found, and the only way in which the weapon could be improved was simply by increasing the size and stiffness of the bow, and the strength of the other components proportionately. This, however, did not increase the muzzle velocity very much, even when a light missile was used, since the inertia of a big, heavy bow would prevent the rapid release of its energy. What it did make possible was the use of much heavier missiles over a comparable range—the heavy bolt and the ball. In the ancient sources there are descriptions of bolts about 6ft 6in (2m) long and about 1½in (3.6cm) diameter. Presumably they had heavy, sharp metal points, and would be fired on a low trajectory, where they could be aimed with some accuracy. They would certainly penetrate any normal body armour, and in many cases the armour shielding (such as layers of hide) used on siege engines. Even a wall or wooden stockade, a perfectly adequate defence against ordinary archery, might be quite useless against them. The other form of missile was a round, or near-round, stone shot. These have been found on ancient sites which are known to have been under siege—or prepared for it—at the time when these machines were in use, and many of them fall within the weight range of 11–55lb (5–25 kg). They would almost certainly be fired, howitzer-style, on a high trajectory to give maximum range and, falling from a height of perhaps 150ft (45m), would make a very powerful impact. Nothing less than a heavy, strong stone structure would afford any shelter, and in an age innocent of explosives the destructive effect must have been immensely impressive. On the other hand, the aim cannot have been very accurate.

The catapults for throwing stone shot were essentially the same as the arrow-shooters, with two slight modifications. On the smaller machines the bowstring was made flat, in the form of a strap, with a loop on its rear side in which the trigger claw engaged. According to Hero, this loop was woven into the fabric of the bowstring. Its height above the *diōstra* was adjusted so that it would thrust on the stone shot centrally, and not ride over the top or slip underneath. On the bigger machines the bowstring was made in two halves, with a leather or hair-cloth sling at the centre, in which the shot was placed before firing, and the ring on the back of the sling was metal. Then, as the bows got bigger and bigger, the windlass was fitted with handles at each end, so that two men could turn it, and was given additional mechanical advantage by the use of a block and tackle. With a 'quintuple-haul' block (see Chapter 4), two men could probably have used a bow with a draw force of 2–3 tons. It may well be asked—did the Greeks of the Hellenistic age ever succeed in making bows on that scale? Here we have a few scraps of evidence from Biton. According to him, one Charon of Magnesia (date unknown, but earlier than second century B.C.) designed a stone-thrower with a *diōstra* about 1ft (30cm) wide. From the size for the sling, Dr Marsden estimated that the stone shot might have been about 4½in (11.5cm) in diameter, and 5lb (2.3kg) in weight. The bow on another machine of comparable size, again according to Biton, was about 9ft

(2.75m) long and 3½in (nearly 9 cm) in diameter at its thickest point. This gives a very rough idea of the general scale and proportions of these machines.

The question arises whether the bigger bows were composite, with or without a sinew layer. Two scientific facts are relevant. The energy-storing capacity of horn is greater than that of wood, and the capacity of sinew is many times greater, so that a simple wooden bow would have been very inferior in performance. Moreover, the energy-storing properties of wood are severely affected by temperature. Above about 25°C (77°F) performance falls off sharply, and such temperatures must often have been reached in the Mediterranean area during the summer campaigning season. The traditional English longbow, made of yew wood, is suited only to the English—or northern French—climate.

In the next chapter, Biton describes an even bigger stone-thrower, designed by Isidorus of Abydos (date likewise unknown). In proportion to the measurements given, this should have had a bow about 15ft (4.6m) long and 1ft (30cm) in diameter, and fired a shot about 9in (23 cm) in diameter, weighing something in the region of 40lb (18kg). Strong evidence that these dimensions are in the realm of fact, and not just an engineer's line of sales-talk, can be seen in the discovery of an ammunition store in Pergamon by the German excavators,* which contained more than 250 stone shot of about this size. One other, really monster catapult deserves at least a mention. According to a near-contemporary account, it was fitted amidships on a very big armed merchant ship built for Hiero II of Syracuse and was designed by Archimedes. It could fire a bolt 18ft (5.5m) long, or a stone weighing 173lb (78kg) up to a range of 200 yards (185m). To score a direct hit on an approaching enemy ship would, of course, take some excellent marksmanship and a bit of luck, but if one did, the stone ball would certainly smash right through the deck and hull of any ordinary ship, and sink it very rapidly.

We must now go back in time, and trace another series of more sophisticated developments which began about halfway through the fourth century B.C. and went on at the same time as the development of the big-bow stone-shooter—namely, catapults with torsion springs. Direct evidence on these comes from the same

*Altertümer von Pergamon (Berlin 1885–1937) X, 48–54, Plate 31.

ancient writers, supported by a few illustrations, and has been supplemented to some extent from another source—experiments carried out on reconstructions. These have taken two contrasting forms. Dr Marsden made, or had made for him, a number of models of various types of catapult, some of them full-scale. He followed the ancient sources as closely as possible, and his reconstructions resembled the catapults in ancient illustrations very closely. He was not, however, able to make authentic sinew–rope (p. 109 below), and the performance of his arrow-shooter, though impressive, fell short of that of the ancient weapons. At the other extreme, a joint project between the Classics and Engineering Departments at Reading University was run during the session 1971–2, and a catapult was designed and built. Owing to budget limitations, authentic materials could not be used, and others, including steel and nylon, were substituted, the design being such that the strength of the structure and the properties of the springs matched the ancient weapons as closely as possible. As a result, the final project had an appearance totally unlike any Greek or Roman catapult, but a performance which ought to be comparable, and some useful information has emerged from the study.

The basic design of the torsion spring was very simple. It consisted of a bundle of strands of elastic material, with an arm or lever thrust through the middle. When this arm was pulled around in a plane perpendicular to the strands, it stretched each of them by an amount proportional to its distance from the axis around which the arm rotated. If the arm was then released, the strands contracted and the arm flew back to its original position. Catapults with one spring mounted horizontally and one arm were in fact a much later development. The standard torsion-spring catapult had two vertical springs, the arms swinging outwards horizontally. Between the tips of the arms was a bowstring, and the rest of the structure was essentially the same as that of a bow-catapult. (Fig. 35.)

Clearly, the most crucial design problem was the choice of material for the strands of the spring, which must meet a number of requirements. Firstly, it must not stretch too easily (in the technical language, the Young's modulus must not be too low), or the movement of the arms, limited to a 90° arc at the maximum, will not generate enough tension. Secondly, it must be a material that can be woven into a rope, since it is very difficult to anchor the ends by any other means. Thirdly, it must be a material which will remain consistently elastic, and not deteriorate under normal weather conditions. And fourthly, ('Which', as the Rev. Mr. Collins said in a different context, 'perhaps I ought to have mentioned earlier') it must not be too difficult to obtain, or too expensive. Needless to say, an ideal material which fully met all these requirements was never found. The nearest approach to it was called, in Greek, *neuron* and the next best was hair.

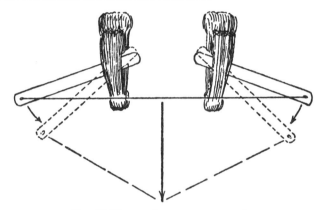

Fig. 35. The torsion spring catapult.

Neuron is usually translated 'animal sinew', but its meaning is not quite so precise as one might wish. It could certainly mean muscle-fibre, or tendon, or a muscle with its tendons taken as a whole. Only in later medical writers (e.g. Galen, second century A.D.) does it come to have its modern meaning—a bundle of nerve-fibres. Hero tells us (*Bel.* p. 110, Wescher) that the best material comes from the shoulders or backs of all animals except pigs, and it has been discovered, he says (indicating some methodical testing) that the *neura* of an animal which get the most exercise are the most springy, such as those from the feet of deer or the necks of oxen. The fact that he says 'feet' rather than 'legs' suggests that he is talking about the Achilles tendon, and not about muscle-fibre, though a tendon is, of course, the termination of a bundle of muscle-fibres, and, as anyone knows who has prepared turkey legs for the oven, the one shades off rather imperceptibly into the other. The regular use of oxen as draught animals with a yoke fitted on their necks, would account for the strengthening of the tendons in that region. A final, clinching argument in favour of tendon is the simple fact that muscle-fibre is edible, and would have gone into the stewpot. Only the tough, inedible, and otherwise useless tendon would have gone into a war weapon.

How was it made into rope? Here, once again, we are faced with an almost complete lack of knowledge of a craft industry. Rope-makers were very much in demand in antiquity, particularly for making ships' tackle. They probably worked for much of the time in the open air, and the average man in the street would have paused to watch them on occasion. As a result, Hero does not feel obliged to say anything about the process, apart from mentioning a 'machine' (or perhaps a tool) for doing the job, without giving any details. It was presumably the same as that for making hempen or papyrus rope, with one difference. The tendon was probably shredded or teased into very thin fibres, each of which would have been shorter than the lengths of other materials such as papyrus. The serviceable length of the Achilles tendon in a large British cow is not much more than 8in (20cm), and almost all the other tendons are shorter. How these strands were woven into a rope which would not creep under very considerable tension is still a mystery.

The Reading University project included a study of the properties of tendon (obtained from a local slaughterhouse) and a stress/extension diagram is shown in Fig. 36. It will be seen that the graph at low strain is not linear, but that beyond a certain point the curve straightens out at a much steeper angle. Other researchers have shown that the point at which this occurs, and the eventual slope, depend on the age of the animal from which the tendon is taken. The accepted explanation for the phenomenon is that the minute fibres from which the material is built up have a 'crimped' structure, and that the behaviour at low strain represents the straightening-out of the crimp, while that on the linear portion is elastic extension of the fibre itself. In their accounts of how to fit a sinew-rope on a catapult, the ancient writers emphasize strongly the need to pre-stress the strands. Though totally ignorant of molecular structures, they must have found by experiment that they could greatly increase the power of the catapult by keeping the sinew under some degree of tension even when the arms were forward and the bowstring straight. Moreover, assessment of the mechanical properties of tendon has shown that they did in fact choose what was probably the best material available to them, since its energy-storing capacity is, believe it or not, higher per unit weight than spring steel. But there is one important qualification which must be borne in mind. All these tests, and the numerous other medical research investigations, have been concerned with individual fibres or very small groups of fibres. When woven into a sinew-rope their behaviour might have been quite different, for two main reasons. Firstly, they are not then extended in a straight line from end to end, but helically, and secondly, their adhesion to each other, which is relatively unimportant in animal tissues (since they are anchored at each end) becomes all-important in the rope.

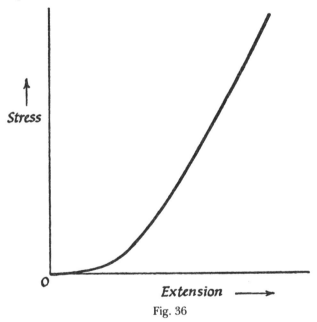

Fig. 36

The other material used for torsion springs was hair. In most contexts where it is mentioned, particularly those in which large quantities are specified, we may assume that horse-hair is meant, but the ancient writers do stress the merits of human hair, especially women's. There were several anecdotes current in the ancient world about the whole female population of cities under siege sacrificing their hair to make catapult springs. One of them concerned the women of Rome in 390 B.C., and was certainly apocryphal, as the Romans had no artillery at that date. This, however, did not prevent later Roman antiquarians from using the story to explain the peculiarly-named cult of the 'Bald Venus' *(Venus Calva)*. Another, told of the women of Carthage in 148 B.C., may well be true.

On the treatment and maintenance of spring materials, as on the making of the rope, we are told virtually nothing in the ancient sources. Hero and Philo both mention that hair and sinew were soaked in olive oil, and treated with grease or fat, both before the rope was made and after it was fitted to the catapult. But Philo points out (Chapter 61) that when sinew-rope is under tension the oil which has been absorbed gets squeezed out, and if more is put on, it will not penetrate. The ropes must therefore be slackened off before oil treatment can succeed. In another writer, a quantity of pine-resin is mentioned in the same context as a consignment of hair for catapult springs, but whether the two items have any direct connection is not certain. If so, the resin, like the oil used on sinew or hair may have served as a water-repellent— Philo notes the destructive effects of rust forming on the iron parts in direct contact with the sinew, and it may also have served as a bonding agent in the making of hair rope. General Schramm, a German army officer who did considerable research on ancient catapults shortly before World War I, successfully used torsion springs made of horsehair, but apparently nobody has yet made a usable sinew-rope (but see p. 226).

The second major problem was the design of the frames around which the rope was to be wound.* To begin with, these were apparently plain rectangular wooden frames with tenons on the ends of the vertical members (which took the stress end-on) and thick bars at the top and bottom. The spring-rope was anchored at one end, then wound around the frame, the other end being tucked under some or all the layers at the top or bottom. At this stage, the rope cannot have been put under much tension except by keeping it taut throughout the winding process, and it is difficult to see how this could have been done. To remedy this, short iron rods *(axonia)* were put under the cords at the top and bottom of the frame and, somehow or other, 'twisted' so as to

*This is a very brief summary of the story; for a fuller account, see Marsden *passim*, and particularly HD, pp. 16–47.

increase the tension. Hero is a bit vague about this, possibly because it was already regarded as an obsolete method by the authors he consulted. The next stage brought a big improvement.

A round hole was bored in each of the horizontal members of the frame, and the spring-rope was threaded through, passing over the iron rod which lay across the hole (Fig. 37) and then back through again. This was repeated until no more strands could be forced through the holes, even using a sort of monster needle in the latter stages. Then the end was anchored as before (tucked

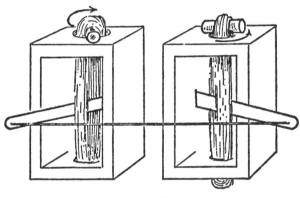

Fig. 37

under the other layers) and the springs were tensioned by rotating the iron rods in the appropriate direction—clockwise for the top left, anti-clockwise for the top right, and so on. This was a very effective method, as it gave a considerable mechanical advantage. But it soon brought out a serious fault in the design. As Hero states (*Bel.* p. 98, Wescher), the most powerful stress in the whole machine acts directly on the iron rod, and pulls it against the wooden frame. Even with very little tension, the friction thus generated would make it extremely difficult to rotate the rods, and even if they could be turned, they would gouge out the wood, especially at the points where they were parallel to the grain.

The simple answer to this problem was to put washers between the rods and the frame to take the wear. On the smaller machines they were round, and made of solid bronze, with a top rim thick enough to support the rods without denting; even so, the area of contact between rod and washer would be less, and the friction between metal and metal much less. For the larger machines bronze washers would have been heavy and expensive, so they were made of wood, with the thrust end-on to the grain, and plates of iron nailed onto the upper and lower surfaces. By this time, if not earlier, a sort of primitive spanner was being used to turn the rods. Hero *(Bel.* p. 101, Wescher) calls it 'an iron lever (literally, crow-bar) with a ring, which engaged with the projecting part of the rod', and it must presumably have looked something like the implement shown in Fig. 38a. Using such a tool, there were obviously two dangers. One was that the washer might shift sideways, causing the outer strands of the spring-rope to rub on the sides of the hole in the frame, producing fraying or abrasion. To prevent it, two small holes could be drilled in the top of the frame close to the hole, and the washer could be made with two projecting lugs which fitted into the holes, and stopped the washer from shifting (Fig. 38b).

The other, more serious danger was that the rods, pressing with great force on the top rims of the metal washers, might produce a sharp burr on their inside edges, which might fray or even shear the sinew rope. The answer to this problem is presented by Hero as an alternative to the fixed washer with lugs; in fact, it was probably a later design which superseded it. The washer was made in the same shape, but with grooves in its top rim in which the rods rested diametrically across the hole, and the washer itself turned round with the rod. This, by the way, would make it easier to use the spanner. To prevent the washer from sliding about sideways, a 'groove' *(solen)* was cut in the top of the frame. This might have been a simple rebate, with the washer shaped to fit into it, or (perhaps) a circular groove, concentric with the spring-hole and of slightly larger diameter, with a projecting rim on the under side of the washer which turned round in it (Fig. 38c & d). This solved the latest problem, but brought back the earlier one of friction between metal and wood, which made the washers difficult to turn, and caused wear on the frame, with eventual weakening. The answer to this, in turn, was to use a round metal 'under-plate' (in Greek, *hypothema*) fixed on the frame, with the 'washer-guide' cut into it. At some stage this seems to have solved the friction problem so well as to raise the opposite one—the washers tended to slip back and loosen off the tension when the adjustment had been made and the spanner was removed. The

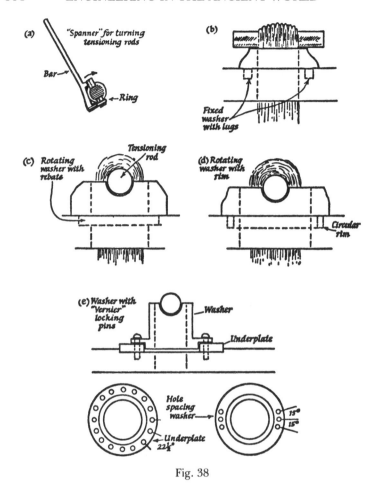

Fig. 38

designers succeeded in solving the problem and another, which has not yet been mentioned, at the same time.

It was clearly necessary to make the thrust of the two springs of a torsion catapult exactly equal. If, for instance, the right-hand one pulled more strongly, the bowstring, and with it the tail of the missile, would be dragged to the right during projection. The groove along which the missile travelled would check this to some extent, but even so, the missile would almost certainly deviate from the line of aim, to the left of the target. The simplest and most obvious way of correcting imbalance (which could be detected by a number of simple tests) would be to tighten up the weaker spring, or slacken off the stronger, by turning the washers, and the engineers seem to have decided that it ought to be possible to make this adjustment accurate to $7\frac{1}{2}°$, or $\frac{1}{48}$ revolution of the washer.

This was the problem, and their solution of it was one of the most ingenious ideas in catapult design. Our evidence for it comes, not from Hero this time, but from archaeology. The remains of an arsenal at Ampurias in Spain (in the extreme N.E., near the modern Figueras) were discovered just before World War I, and the finds, to be dated about mid second century B.C., included most of the metal parts of a catapult. Among them were washers and 'underplates' of a new design. The washers were shaped as shown in Fig. 38e, with a flange which rested on the raised part of the 'underplate'. The slip-back was prevented, very simply, by two pins, diametrically opposite each other, inserted through holes in the flange and in the underplate and—for a short distance at least—into the wood below it. Now in order to get an adjustment to the required accuracy, it might seem necessary to drill two holes in the washer flange and 48 holes around the rim of the 'underplate' at intervals of $7\frac{1}{2}°$, leaving only very thin walls between the holes and, inevitably, weakening the 'underplate' very seriously. But this was avoided by what might be described as a very primitive Vernier system. 16 holes were drilled in the 'underplate' at a spacing of $22\frac{1}{2}°$, and three pairs of diametrically opposite holes in the washer at a narrower spacing—15°. Thus at every step of $7\frac{1}{2}°$—i.e. at 48 points per complete revolution—one or other of the three pairs of holes in the washer would line up with one of the eight pairs of holes in the 'underplate', and the washer could thus be pinned in any one of 48 positions. Bearing in mind that, to judge from illustrations, the washers were quite often given as much as three complete turns during the tensioning process, the addition or subtraction of $\frac{1}{48}$ turn would have been quite a fine adjustment—roughly $\pm 0.5\%$.

Simultaneously with the development of the washers and tensioning apparatus came various improvements in the design of the frames. To begin with, as already mentioned, the two spring-frames were made separately, and attached to the sides of the 'trough'. But this was a bad arrangement, as drawing back the bowstring puts a very strong torque on each frame, tending to rip it off its fixing at the front. Various remedies were tried, including planks to join the two frames, but they increased the weight of the machine without improving its power, and made it more difficult to dismantle and re-assemble—a requirement stressed by the ancient writers. After a short period of thought, the design was simplified by the use of a single length of wood to form the tops of both frames, and another for the bottoms. The stretching of the springs then tended to push the whole frame, as a single unit backwards along the 'trough', but it could be fixed by wooden dowels, or perhaps removable pegs.

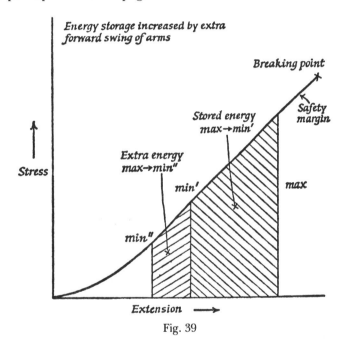

Fig. 39

The next concern of the designers was to increase the arc over which the arms could move. To get the maximum energy storage and release, it was necessary to make the difference between maximum strain (with the arms drawn back) and minimum (with the bowstring straight) as wide as possible, and one way of doing this was to allow the arms to swing further forwards.*

*The energy is represented in a graph of stress/extension (Fig. 39) by the area below the curve, between the vertical lines representing

Closely related to this was a problem which had already been encountered. Though the tension in the springs had fallen off, the arms were still moving quite rapidly at the end of their swing, and carried considerable kinetic energy which had somehow to be absorbed. It was soon found that if the arms struck the outer uprights of the frame they would be severely damaged. On the other hand, if the bowstring was made shorter, so that it pulled the arms up before they reached the frame, it would have to absorb their energy, and would be suddenly and very violently stretched. Machines adjusted in that way must have needed frequent replacement, both of the bowstring and of the arms which, when the bowstring snapped, would smash themselves on the frame uprights. Philo remarks (Chapter 68) that breakage of an arm was one of the commonest accidents to befall catapults: it not only put the weapon out of action for a long time, but was also highly dangerous to the crew, as the broken portion flailed across in front of them. This happened during a deliberate overload test on the Reading project catapult, and amply justified the rigorous safety precautions which had been enforced.

The answer to the problem came when the designers realized that the torsion springs could, in another mode of operation, also be made to serve as shock-absorbers, and take up some of the energy which would otherwise have gone into the bowstring. This was done by extending the 'butt ends' of the arms (which they called 'heels', *pternai*), reinforcing them, and positioning pads on the inner frame uprights so that the 'heels' struck them just before the bowstring straightened. From then on, the 'heel' would act as a fulcrum, and some of the energy from the moving arm would be used up in displacing the middle of the torsion spring sideways.

As a result of this modification, the heel-pad became the limiting factor on the forward swing of the arm. In order to increase that swing, the inner upright was shifted backwards (i.e. away from the 'heel'), and the outer upright forwards. In order to give room for this, while at the same time economizing on material, the top

---

maximum and minimum extension. The maximum can be pushed up (i.e. to the right) provided it falls short of the breaking point by a reasonable safety margin. If the minimum is then lowered (i.e. to the left) the area (which represents stored energy) is increased, though this has progressively less effect as the stress falls to lower values.

and bottom members of the frame were made with a convex rear edge. Since the heel-pads required a certain amount of 'give', the earlier problem recurred of the arms striking on the outer uprights, and so a recess was cut out from the centre of their rear edges, just wide and deep enough to clear the arm as it swung forward. To compensate (partially) for the loss of strength which this caused, the front edge of the upright was made convex (Fig. 40).

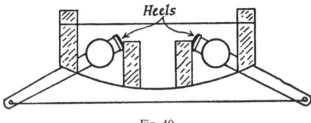

Fig. 40

This design came to be regarded as the optimum for a moderately light, high-velocity arrow-shooter. It was known as 'straight-spring' (in Greek, *euthytonos*) from its general appearance (it should be remembered that the off-setting of the uprights is not visible from above). They found, however, that in order to throw heavier missiles it was necessary not only to increase the size of the springs, but also to extend the forward swing of the arms. This was done by carrying the same method of off-setting still further, until eventually the front edge of the inner upright was made level with the rear edge of the outer. The geometry of the design, as given by Hero (*Bel.* p. 94, Wescher) is as follows: (Fig. 41).

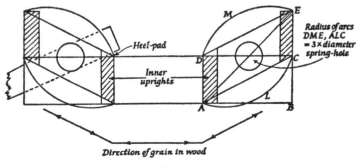

Fig. 41

Though the difference between this design and the 'straight-spring' is one of degree rather than kind, the term used for it reflected precisely that difference and no other: it was called a 'fold-back spring' *(palintonos)*. In practice, however, the much more obvious difference lay in the fact that the 'straight-spring' was normally an arrow-shooter, and the 'fold-back spring', almost by definition, a heavier and more powerful stone-thrower.

There is just one peculiar feature. Hero seems to imply that for the *palintonos* they reverted to the old system of making the two spring-frames separate, and used an array of planks and struts to fix them to each other and to the 'trough'. He gives no reason for this, but a simple explanation can be offered. The portion of the diagram marked ALCEMD has to withstand the greatest stress in the whole machine. If the grain of the wood ran parallel to DC its load-bearing capabilities would be very limited, because almost all the stress would fall on the cross-bonding between the fibres, hardly any of which would be effectively supported at both ends. It would not take much tension in the spring to split the wood along its grain. If, however, the grain ran parallel to AC and DE, a fair proportion of its cross-section would be supported at both ends. (The grain of the left-hand frame would have to slant in the opposite direction.) As it was patently impossible to find lengths of wood in which the grain had two bends of about 26° at the right places, they used three separate pieces—hence the separate frames. Why did Hero not explain this? Probably it was left the carpenter to point out what was necessary, and why ('if you puts 'un on the slant to the grain, 'twill jigger 'un up, m'dear'). Such vital steps in the development are not recorded in our sources.

A very brief mention will have to suffice for the other details of these two types of catapult. Each was fitted on a tripod base, the attachment (at or near the centre of gravity) being by a swivel mounting (in Greek, *carchesion*, see p. 95) which allowed the whole weapon to be tilted up or down and to swing sideways into any position. The bigger stone-throwers had a beam-and-strut structure to support the 'trough', instead of a solid batten, which would have been too heavy. Another feature very much emphasized by all the ancient writers is that those portions of the machine, particularly around the frame, which were subject to shock or wear had iron plates nailed over them.

Something has already been said about the dimensions of bow-catapults, but on the torsion-spring types we have considerably more information, thanks mainly to Philo and Vitruvius. These catapults were classified in terms of the missile they were designed to shoot—the weight of the stone shot or the length of the bolt. The latter may sound rather crude, but it is perfectly clear that arrow patterns were strictly standardized, and that for any given length there was an accepted diameter and weight. This applied equally to the very large bolts. Thus we hear of a 'three-span *euthytonos*'—which was a straight-spring model firing a bolt about 27 in (0.68 m) long, or a 'twenty-mina *palintonos*'—a 'foldback' spring model throwing a stone shot of about 19lb (8.6kg).

The crucial dimension, as the ancient designers well knew, is the size, and hence the power, of the springs. A theoretician might measure this in terms of the weight of sinew-rope, but in the practical directions given by Hero, Philo and Vitruvius each spring is regarded as a cylinder of set proportions, and the crucial measurement is its diameter, which is taken to be the same as that of the holes in the tops and bottoms of the frames through which the spring-cord was threaded. All the authorities agree on two formulae, one for each type of catapult. For a straight-spring machine, the diameter D is $\frac{1}{9}$th of the length of the bolt, and for the *palintonos* (stone-thrower) $D = 1.1\sqrt[3]{M} \times 100$, where *D is* measured in *dactyls* and *M* is the weight of the shot in Attic *minas.* * Thus, for a three-span arrow-shooter the diameter *D* would be $\frac{36}{9} = 4$ *dactyls*, just over 3in (7.72cm), and for a twenty-mina stone-thrower, just over $10\frac{1}{2}$in (26.75cm).

Clearly, the first of these formulae does not involve any arithmetical difficulty, but the second, without four-figure tables or a slide-rule, is not so easy. It is simple to work out if *M* happens to be 10 or 80 *minas,* since 1000 and 8000 happen to be cube numbers, but what could the ancient artificers do when *M* = 15, or *M* = 45? Three possibilities are mentioned by the ancient authorities. Some workshops were lucky enough to have lists of measurements for the various sizes of shot, worked out by a tame

*The Greek *dactyl* was very nearly $\frac{3}{4}$inch (0.76 in) or 19.3 mm. A 'span' (*spithamē*) was 12 *dactyls* and a 'cubit' *(pechys)* 24 *dactyls*—roughly 9 in and 18 in (23 and 46 cm) respectively. The Attic *mina* (there were other standards in use) was just under 1 lb (0.96 lb) or 436 g. One 'talent' *(talanton)* was 60 *minas,* roughly 58 lb or 26 kg.

mathematician. The figures given by Philo and Vitruvius were derived from such lists. Failing this, there is the cruder device of finding (by trial and error) the nearest whole number to the cube root: for example, if $M = 45$, $100 \times M = 4500$, and the cube of 16 is 4096 and that of 17 is 4913. Having found this, they could either split the difference or take 16 or 17 as though it were the exact cube root, from motives which might be economic, or personal, or even dishonest.

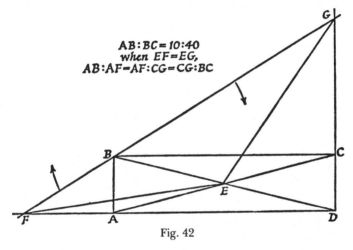

AB:BC = 10:40
when EF = EG,
AB:AF = AF:CG = CG:BC

Fig. 42

The third method, described by Hero and Philo, is much more accurate. (Hero, *Bel.* p. 112–119 Wescher, Philo, Chapter 51–2). It starts from the known spring-diameter of a successful catapult. For this purpose, let us take a ten-mina machine, where $M' \times 100 = 1000$, and $D = 11$ *dactyls*. It is possible to find the correct diameter for a second machine (larger or smaller) by calculating two mean proportionals (in Greek, *mesai ana logon*) between its missile weight M'' and that of the first machine, M' (10 minas). Suppose $M'' = 40$ minas, we then require two numbers, x and y, such that $10 : x = x : y = y : 40$. The simplest way of finding these is by a geometrical construction, which would be within the scope of all but the very dimmest artificer. This version is Hero's, and Philo's differs only very slightly in detail. We draw two lines AB and BC at right-angles, their lengths proportionate to M' and M'' (i.e. in the ratio 10:40, Fig. 42). Complete the rectangle ABCD, and draw its diagonals AC, BD, intersecting at E.

Produce DC and DA for some distance. Then place a straight-edge, pivoting on B and cutting the extensions of DA and DC. Rock the straight-edge about B until each of these points of inter-section is the same distance from E, measured with a ruler, and call these points F and G (i.e. FE = EG). It will then be found that AF and CG are the mean proportionals between AB and BC—that is, AB : AF = AF : CG = CG : BC. It also follows that the ratio AF/AB is the cube root of the ratio BC/AB, and that the spring diameter suitable for M' (11 *dactyls*) multiplied by this ratio (AF/AB) will give the correct spring-diameter for M''.

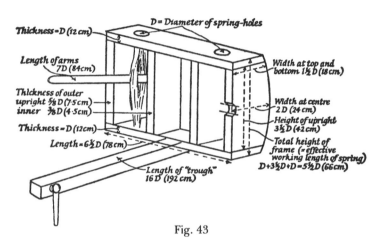

Fig. 43

Once this basic dimension D has been calculated, the rest be-comes easy. Both Philo and Vitruvius give a long list of other dimensions, all expressed in terms of this diameter. Some minor differences between their figures can easily be accounted for by the developments one would expect to take place in the gap of nearly two centuries which separates them. Here (Fig. 43) are a few of the typical figures for a *euthytonos* ($D = \frac{1}{9}$ of the length of the arrow). Let us take D = 12cm. A corresponding set of dimensions for the stone-thrower *(palintonos)* shows a few predict-able differences. The arms are shorter (6D), but had a wider arc of movement, so the 'trough' is longer (19D). The tops and bottoms of the arrow-shooter's two frames are each made from a single piece of wood, and its length is $6\frac{1}{2}D$. The corresponding meas-urement for the *palintonos* is $2\frac{3}{4}D$, which is consistent with the supposition (p. 119) that two separate frames were used on this machine. Allowing $1\frac{1}{4}$ D for the width of the 'trough' between them, the total width would be $6\frac{3}{4}D$, a little more than the *euthytonos*.

In conclusion, a brief account must be given of six developments in catapult design, four of them Greek and two Roman. The four Greek ones—the repeater, the wedge, the bronze-spring and the pneumatic—are all described by Philo. The first two may possibly have been developed as working weapons, but the bronzes-pring and the pneumatic were almost certainly not.

The employment of the arrow-shooting catapult in ancient times corresponded quite closely with that of the machine-gun in modern warfare, up to the time (towards the end of World War I) when a really effective answer to the static machine-gun post was developed, namely, the tank. The main function of the catapult was to pin down enemy infantry. Troops approaching a fortified position had either to risk heavy casualties or else move in small groups, keeping to the 'dead ground' as much as possible, and if the defenders' catapults had been skilfully positioned, this might be very difficult. Conversely, the attackers could use their own catapults to force the defenders to keep down behind the cover of walls or battlements, where they could not take such effective steps to prevent the scaling of the walls or the use of siege-engines. In each of these roles, the speed of fire was clearly an important factor in the effectiveness of a catapult, and it is not surprising that some attempt was made to convert the simple machine into what nowadays might loosely be called an automatic weapon. In Greek it was called a *polybolos*—'multi-shooter'. It was developed in Rhodes by one Dionysius, who came originally from Alexandria. From this we may infer that there was some degree of 'brain-drain' going on between the various armament centres in the Mediterranean area, and it is amusing to see how Philo, who travelled around several of them, was on some-occasions regarded with suspicion, and perhaps deliberately misled by security-conscious technicians.

The essential steps in firing a catapult which, it would appear, were normally carried out by a crew of at least two, were as follows: (1) push the slider *(diōstra)* forwards (2) lock the 'claw' on the bowstring (3) draw back the *diōstra* (4) place the missile in the groove (5) take aim (6) pull the trigger. Dionysius succeeded in automating (2) (4) and (6), and the business of aiming was effectively eliminated by pre-aligning the catapult on a tripod and firing it on fixed lines. The not-too-difficult task of sliding the *diōstra* back and forth was speeded up by a chain-drive, so that one man, with practically no training or experience, could use the weapon.

The chain-drive deserves special mention, as it has possible applications in other fields (see p. 73). This is how Philo describes it (Chapter 75–6). 'It (i.e. the *polybolos*) does not have a cord for pulling back the *diōstra*, and instead, the capstan is made with five facets on each of its projecting ends. Small blocks of oak (literally, "briquettes", *plinthidia*) with iron plates on each side are mortised into each other and linked with pins, and these are wrapped around the capstan. There are (chains of) these on either side of the "trough", and each chain is fixed to the *diōstra* (at one point) by a pin with a round head.' The arrangement was as shown in Fig. 44a. Philo mentions handspikes (in the plural) on the capstan; once again, it is surprising that a crank was not used.

The trigger, instead of a simple lever (see Fig. 44b) was in the form of a rocker-arm, activated by vertical studs at one side of the 'trough', so that its working end was pushed under the back end of the 'claw' when the *diōstra* was fully forward, and was pulled away to release the 'claw' just after the arrow had fallen into place and the *diōstra* reached its backward limit. The ammunition feed was an ingenious mechanism, consisting of a long wooden cylinder (something like a rolling-pin) which revolved inside a wooden tube with slots at the top and bottom just wide enough for a bolt to fall through. Over the slot at the top was fixed the magazine, big enough to hold a considerable number of bolts, with sides sloping inwards towards the slot (Fig. 44c). The cylinder itself had a groove running from end to end, also just deep and wide enough to take a bolt. When the cylinder was turned so that the groove lined up with the top slot, a bolt would fall through into the groove. It was then rotated half a turn, so that the groove came in line with the lower slot and the bolt fell through onto the *diōstra*, just in front of the bow-string. The lower slot had just enough clearance above the *diōstra* for the trigger mechanism to pass under it. Philo is a little confused about the bolts (or so it would seem). He says, quite rightly, that they could not be notched, since they might not happen to drop in the right position, but he also says that they had

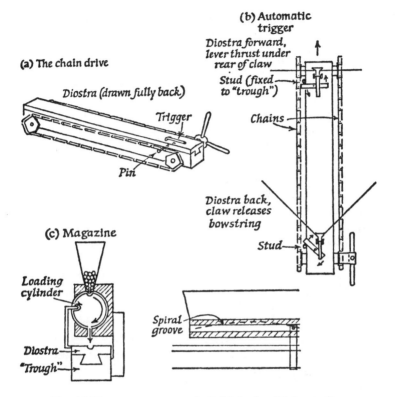

Fig. 44. The repeater catapult *(polybolos,* 'multi-shooter').

three flights (i.e. they were arrows, not bolts—the same Greek word is used for both). Even if this means that they had three flights one behind the other and all in line, it is difficult to see how they could pass through the loading-cylinder or lie flat in the groove on the *diōstra,* if they did not happen to fall exactly right. The rotation of the cylinder was also automatic. One side of the tube was partly open, and the cylinder had a spiral groove on that side, which ran halfway around the circumference. A short bar with a knob on its end was fixed on the *diōstra* to engage in this groove, and this turned the cylinder so that the bolt-groove faced upwards when the *diōstra* was fully forward, and downwards when it was almost fully back.

Having described this machine, Philo goes on to make some disparaging comments on it. His motives for doing so are not clear, unless perhaps he could not remember some of the details accurately (occasional remarks suggest this, and at one point he actually admits it) and he might have been embarrassed if called upon to construct one. His main criticism is that it fires along fixed lines, and is therefore useless against a moving target. Even firing at a group might not be effective, because the beaten zone is short and narrow. But this criticism would not apply when the catapult was laid (for instance) on a crucial, narrow approach to a fortified position. There are, moreover, two considerations which would seem to override the objection. Firstly, like the Bren gun mounted on a tripod, the catapult could be set up and aimed at an enemy camp during daylight, and used in the darkness with devastating effect. Secondly, experienced men who had been under fire from ordinary catapults would be able to gauge with some accuracy the time it took to wind back the slider and get ready to fire the next bolt, but when encountering this type for the first time, they would greatly over-estimate the 'safe interval' after each shot, and might fall easy victims when they thought they were safe. However, Philo does not seem to have appreciated this. His great worry, as an artillery expert, is that unskilled infantrymen might get hold of the weapon, blaze away indiscriminately, and waste ammunition.

The second of the Greek developments was an alternative method of tensioning the springs on a torsion catapult. Philo himself may have been involved in the research, since he certainly spends a great deal of time criticising the washer method (p. 112) and expounding what he considers to be a great improvement, (Chapter 56–67). What it all boils down to is that he returned to the original, simple frame without spring-holes. Along the tops and bottoms of the frames were semi-cylindrical bars, to distribute the pressure and wear over as much of the sinew-rope as possible. Then the sinew-rope was wound around the outside of the frames, and not pre-tensioned—or only a very little. Next, wedges were driven in between the top members of the frames and the semi-cylindrical bars, in effect expanding the frames and stretching the springs. The two obvious advantages are (a) that for a given size of frame more sinew-rope can be wound on, and (b) that the bundles of sinew-rope are further away from the axis around which the arms rotate, and so the leverage is greater. Philo claims a number of other advantages, but he is not altogether objective in discussing what might have been his own brain-child.

The third development was the bronze-spring catapult, said to be the invention of Ctesibius. Philo describes its design rather cursorily, and adds a number of misguided comments on the underlying theory. Though he compares the properties of the bronze used, in a rather confusing way, with those of 'so-called Celtic and Spanish swords' which were made of steel, there is no doubt that the name 'bronze-spring' (in Greek, *chalcotonos*) was an exact term, and did not merely mean 'metal'. The alloy used had a slightly higher tin content than usual—'three *drachmas* of tin', he says, 'were added to each *mina* of best quality purified red

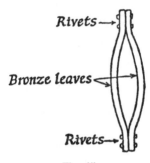

Fig. 45

bronze'. A *drachma* is (rather conveniently) one-hundredth part of a *mina,* so this would increase the tin content by 3%. Philo's readers could be expected to know exactly what the normal tin content of 'red bronze' was, but we unfortunately do not. Great care was taken to obtain the best quality copper and tin, and to purify them by repeated smelting, and the alloy was cast in the form of two rectangular strips. Philo gave the exact measurements, but the figures have not been preserved in our copies of the manuscripts. They were hammered to the right thickness, bent around a wooden roller, and then cold-forged for a long period, using light hammers and light blows. According to Philo, the effect of this was to case-harden them. If heavy hammers were used, the springs would be hardened right through, and would become too brittle. Then their ends were shaped and filed, and riveted together to form an elliptical leaf spring (Fig. 45). One of these was mounted on each of the outer uprights of an ordinary catapult frame (a makeshift arrangement, which suggests an early stage of research), and held in iron brackets, which also held the axles on which the arms turned. A small projection on the 'heel' of each arm (Philo calls it a 'bronze finger') thrust against the spring and compressed it as the bowstring was drawn back (Fig. 46). Philo then describes (with disapproval based on a completely faulty theory) how Ctesibius attempted to increase the power by mounting two such springs side by side on each arm.*

Philo claims, quite rightly, that bronze springs are less susceptible to damage or deterioration through bad weather conditions, but he also claims that they were more powerful than sinew-cord, which is most improbable. The whole design has the appearance of a tentative research project, and there is no evidence to show that catapults of this type were ever in general use.

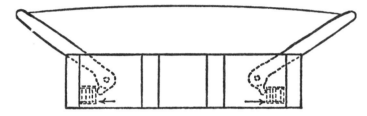

Fig. 46. Bronze-spring catapult.

The fourth Hellenistic Greek development was the pneumatic catapult (in Greek, *aerotonos*, 'air-spring'), also attributed to Ctesibius. The description of this comes right at the end of Philo's work, and is included, he says, 'so that nobody may think that my researches are incomplete'—an interesting and typical intrusion of the public-relations angle. Whereas his account of the making of bronze springs is in the first person ('We hammered the plates . . . etc.'), that of pneumatic catapult is at second hand ('We were told how Ctesibius demonstrated. . . .'). Clearly, we are dealing with a tentative research project, which had apparently been abandoned by Philo's time—about 50 years later.

The compressibility and elasticity of air had been known and understood from quite early times, and a theoretical basis of a sort was established by Strato of Lampsacus (third century B.C.). But all ancient authorities seem to agree that Ctesibius was the

*Marsden's interpretation of this passage (*T.T.* p. 144) is incorrect, and adds further confusion to Philo's already confused thought.

first to make practical use of it in a number of mechanical devices, of which the pneumatic catapult was the most impressive. Another was the water-organ *(hydraulis)*, and Philo counters possible scepticism in his readers by pointing out that the piston-and-cylinder pump on that apparatus, which they could see for themselves, worked quite effectively.

The layout of the catapult was the same as for the bronze-spring machine, except that pistons and cylinders (with no outlets) were substituted for the bronze plates. As the arms were drawn back, some sort of tubular projections on their 'heels' forced the pistons into the cylinders, and compressed the air in them. When the bowstring was released, the pistons popped outwards and swung the arms forwards. The pistons and cylinders were made of bronze, initially cast, and then hammered on the outside. The cylinder, roughly shaped in the casting, was bored out with some precision, and the piston turned to fit it. Unfortunately, Philo assumes that his readers know what tools were used and how, and gives no details.

He does tell us, however, of a demonstration given by Ctesibius, and reported by his informants, of the pneumatic 'spring'. The cylinder was put into a clamp or vice, and the piston was driven into it with a hammer and wedge. After a time, the air pressure built up so high that even a firm blow from the hammer would not force the piston in any further. When the wedge was knocked out (sideways?) the piston flew out with great force (and, we may assume, a loud bang). The onlookers even reported that a flame shot out with the air, a phenomenon which they explained as the effect of air rushing over the metal surfaces at high speed and being heated by the friction. This explanation is a very familiar one, being used by the philosophers as early as the fifth century B.C. to account for lightning, and even if Philo's informants imagined the whole thing, or made it up in order to impress him, it is a fact that the cylinder would have heated up considerably as a result of the air being compressed inside it.

It seems almost certain that this catapult never got beyond the experimental stage. The reasons are obvious; to bore the cylinder and turn the piston (a piston without rings, and with no gland on its driving-rod) with sufficient accuracy was beyond the technology of the time. If the piston fits too tightly, friction will prevent it from moving fast enough, and if it is not tight enough, the air pressure will leak away. The best compromise between these faults that could be reached in Alexandria in the third century B.C. was not good enough, though surviving pumps from not much later show that their skill was by no means negligible.

Finally, two Roman achievements—the *cheiroballistra* ('hand-catapult') and the *onager* ('wild ass'). Our evidence for the first of these comes from the incomplete and rather problematic remains of a treatise of that name by Hero, and from some illustrations, notably those from Trajan's column. It was, as its name suggests, a compact, portable arrow-shooter, with frames that were made entirely of iron. Each torsion spring had two bars (corresponding to the outer and inner uprights of the wooden frame) with rings at the top and bottom in which the washers and tensioning rods were fitted. They were held in position by two cross-members. The lower one (to which the 'trough' was fixed) was called the 'little ladder', being of a beam-and-strut construction, and the upper one the 'little arch', because it had an inverted U-bend in the middle. The fixing of these components must have been strong and rigid, though how it was done is not clear. Dr. Marsden suggested welding, but that must have been quite difficult in a Roman forge, without modern equipment. Riveting is an alternative, and was used on Marsden's full-scale model (see *T.T.* pp. 232–3 and Plates 6–7–8). The 'trough' and slider were of wood, and the weapon was otherwise exactly like the older versions.

Its advantages over the older models were considerable. Though probably not much lighter in weight, it was more compact, and easier to set up—some illustrations show a semi-mobile version, fitted in a small cart. The sinew-springs were protected by metal cases from enemy weapons and from the weather. In addition, they were set further apart, which with the shape of the 'little arch', gave a wider field of view and made aiming easier. The fixing of the iron frames was so managed that the arms could traverse a wider arc than those of a wooden-frame machine, thus providing extra power and range. In this instance we have good evidence (particularly from Trajan's Column) that the weapon was fully developed and successfully used.

All the catapults so far described had two vertical springs, and two arms which swung horizontally. What, then, of the *onager,* that machine familiar from many cartoon drawings and classroom models, with one horizontal spring and one arm which swings upwards? There is remarkably little evidence for it in classical writers. None of the main treatises mentions it (Hero, Philo, Vitruvius), and the earliest account which gives any detail at all (and, by comparison with Hero and Philo, miserably little detail) comes in the writings of Ammianus Marcellinus, who lived from about A.D. 330–390. The surviving part of his history deals with the years 353–378 A.D.

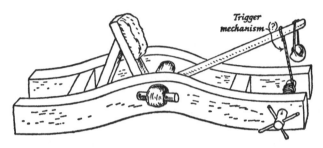

Fig. 47. The *onager*

The gist of his account says that the *onager* had two main horizontal beams running from front to rear, made of oak or ilex, which 'rose in a hump' near the middle, and were held apart by cross-beams. A large hole was bored in each of the main beams, through which 'the ropes' passed—we are left to guess that they were sinew-ropes, tensioned with washers and bars in the same way as the earlier machines. A wooden arm was thrust into the middle of the bundle of ropes, with iron hooks on its tip to take the sling, made of tow or (rather surprisingly) of iron. As there was no bowstring, and the arm apparently had no 'heel', some other method of stopping it at the end of its swing and absorbing its kinetic energy had to be devised. This was done by mounting a large cushion, stuffed with chaff, on a braced structure above, and just in front of, the spring (Fig. 47). The arm was winched down by two or more men, using a windlass at the rear end. Then the shot was placed in the sling, and the sergeant in charge *(magister)* 'drove out the bolt which retains the whole rope-mechanism by a mighty blow with a hammer'. It is a pity that Ammianus could not bring himself to cut down on the drama and give us more technical details of the trigger mechanism.

From the silence of earlier authors we may infer that the *onager* did not come into general use until much later than the two-spring catapults, and from Ammianus that it reached its greatest popularity in the third or fourth century A.D. Why should this be so? It was probably a good deal easier and simpler to make than the two-spring stone-thrower, and it did not require so much maintenance or adjustments such as the balancing of the springs, or accurate alignment of the slider and trough. So long as the available technical skills were adequate for this, the two-spring machine had the obvious advantage of better performance, but in later days, when craftsmanship declined, the simpler and cruder machine became preferable. In order to give a comparable range, the single spring of the *onager* would have to be made very large. On occasion it was apparently enormous—Ammianus speaks of eight men being needed to wind down the arm, even with a windlass. He also says that the recoil was tremendous (meaning the reaction on release, which would thrust downwards on the rear end of the machine, and might even cause the front end to jump up)—so much so that it could not be positioned on top of a masonry wall, because it would dislodge stones, 'not by its weight but by the violent impact *(concussione violenta)*'. It was, he says, used on a turf platform, which could absorb the shock.

The performance of the *onager* was probably improved quite considerably by the use of the sling, and some interesting results might come from more detailed research into the trajectory of the missile with various lengths of sling and different adjustments of of the hooks. This may well have been the method used to alter the range—a longer sling gives a lower, flatter trajectory and shorter range—but it is difficult to see how the direction of aim could be altered, in view of the fact that the whole machine would have to be shifted round. A full-scale model* weighed over two tons, and many of the ancient machines must have been heavier still. Only a very modest-sized one could possibly have been mounted on wheels or made effectively mobile.

*Described in the Appendix to *The Crossbow* by R. W. F. Payne-Gallwey (2nd ed., London 1958).

6

*Ships and sea transport**

FROM very early times the lives of the Greek people have been closely linked with the sea. Many of their city sites were chosen mainly, if not entirely, because a good anchorage was available nearby. Much of their import and export trade was carried by sea, overland transport being slow and costly. It was naval supremacy in the Aegean that enabled Athens to dominate that area for the greater part of the fifth century B.C., and to make various communities, who had joined her as fellow members of a defensive alliance, into the subjects of an Athenian empire. That same supremacy also enabled her to operate protected shipping routes, and import large quantities of cereals from the Black Sea area and elsewhere. This was especially important during the Peloponnesian War, when Spartan invasions cut off access overland and destroyed the local crops before they could be harvested, but importation remained vital during the following century and later.

Preoccupation with the sea and ships often reveals itself in the imagery used by Greek poets. The 'ship of state', with a political leader as 'helmsman', became a cliché, and when Haemon (in Sophocles' *Antigone,* lines 715–7) tells his father that 'a man who keeps the sheets taut and refuses to slacken them in a squall will finish his voyage keel uppermost', his words must have had a vividness and immediacy for the audience which they cannot have for many of us today.

There were also, inevitably, important consequences for technology. For any city-state to maintain her dominance, it was necessary for her ships to be superior not only in numbers but also in design and performance. The principal technique of naval fighting at that time was ramming, and its whole success depended

*For the sake of simplicity all measurements in this chapter are expressed in modern units: a table of Greek and Roman units and their equivalents is given in Table 3 on page 169.

on speed and manoeuvrability; even a very slight advantage in either could make the difference between victory and defeat. So, when city-states were willing to make financial resources available and kept demanding new and better ships, it is not surprising that a considerable body of expert knowledge was built up in the Greek shipyards, or that ship design attained a high level of competence and success.

The Roman approach was very different. At no stage in their history were they really a seafaring people. Their capital city was sixteen miles inland, with access to the sea (via the river Tiber) only for small boats, and Ostia, at the river mouth, was quite insignificant as a sea port until the major harbour works were completed by the emperor Claudius in the mid-first century A.D., by which time Rome had become dependent on the importation of grain from overseas. As for naval strength, it was not until the start of the Punic Wars in 264 B.C., when the Romans had to face the Carthaginians—a nation with strong naval forces and a seafaring tradition—that they built their first fleet of warships. For several centuries before that they had managed without one. It says much for their courage and determination that they were willing at that time to learn the task of building ships and training crews from an army of landsmen.

Even so, at the end of the Second Punic War (201 B.C.) when the Carthaginian naval strength was finally broken, the Romans allowed their fleet to deteriorate without renewal, and when fleets were needed in later times they were either raised from the Greek cities in southern Italy, or specially built. It was not until after the Battle of Actium (31 B.C.) that the first permanent fleet was organized by Augustus, and permanent Roman naval supremacy in the Mediterranean was established.

The Romans contributed little to the development of ship design. They took over where the Hellenistic Greeks had left off, with a preference for a type of warship which had been in existence for more than a century. Their merchant ships also seem to have been very much like those of their Greek predecessors. It is by mere chance that a greater number of Roman illustrations have been preserved.

An idea of the form of an early Greek ship can be gained from two pieces of evidence—a passage from Homer's *Odyssey* (V, 228–61) and a cup by the vase-painter Exekias dated to the late sixth century B.C.* Fig. 48 is based on the cup illustration, with some details added from other sources. It shows a pirate boat which could be rowed or sailed; Odysseus' boat was for sailing only.

The fact that each of these sources gives any kind of technical information is in itself surprising. Odysseus has been detained on an island by the beautiful nymph Calypso for a number of years.

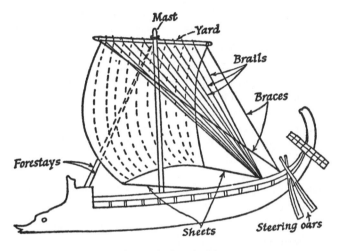

Fig. 48. Early Greek ship

When he is at last allowed to leave she does not conjure up a boat by magic or produce one which she has kept hidden. Instead, she presents Odysseus with a set of tools, and shows him where to find suitable timber on her island. Odysseus turns out to be not merely the warrior hero of epic tradition, but also a skilled and knowledgeable craftsman. Homer clearly assumed that his audience knew about, and were interested in, the details of shipbuilding and equipment, and Exekias, though he was illustrating a fantasy scene from mythology, has taken care to draw the vessel correctly down to quite minor detail.

In this chapter, oared vessels will be discussed first, and sailing vessels later. But first, an important difference between all types of Greek and Roman boats and modern ones should be noted, which

*Arias-Hirmer-Shefton, *A History of Greek Vase Painting* (Thames & Hudson, London 1962), plate XVI.

is not obvious at first sight from the illustrations, but is made quite clear by the Homeric account and from archaeological evidence. It concerns the method of hull construction.

In recent times, most wooden hulls have been made by a method called 'clinker' or 'lapstrake'. The keel is first laid down, with a heavy, shaped beam ('keelson') on top of it, and the stempost and sternpost jointed into it. Then vertical frames are attached at intervals, which determine the eventual shape of the hull. The outside shell of planks is then put on, starting with one either side of the keelson ('garboards') each overlapping the one below by a small amount. To keep these overlapping joints tightly together, it is necessary for the planks to run effectively in one piece the full length of the boat, and to be suitably springy.

This method of construction is characteristic of Northern Europe, particularly Scandinavia, though the time and place of its origin are obscure. The older method used by Greeks and Romans, and in the Middle East until quite recent times, belonged to the Eastern Mediterranean and may have originated in Egypt. It is known as 'carvel' construction.

Instead of overlapping, the planks of the outer shell are jointed edge-to-edge. In very early times they were held together by ropes or threads passing back and forth through holes—as the Greeks said, 'stitched together'. Ancient commentators on Homer, and some modern scholars, have thought that he refers to this when he makes Agamemnon warn the Greeks (in *Iliad* II, 135) that, after some years on the beaches of Troy 'the timbers of the ships are rotting and the ropes losing their tension'. If so, this must be a deliberate archaism, since Homer makes Odysseus build his boat by a much more sophisticated method. Perhaps Agamemnon was just referring to the ropes in general.

The later method of edge-jointing the planks, described in *Odyssey* V, 246–51, involved a fair amount of joinery such as could only be done by a skilled carpenter. At intervals of 3–9in (7–22cm) depending on the size of the ship mortises were cut into each of the facing edges exactly opposite each other (Fig. 49) and short wooden tenons made to fit into them and form a bond across the join. These were cut with their grain end-to-end, at right-angles to that of the planks. Their thickness was usually $\frac{1}{3}$ to $\frac{1}{2}$ that of the planks, and their width varied according to the size of the ship and the distance between them. When Homer says

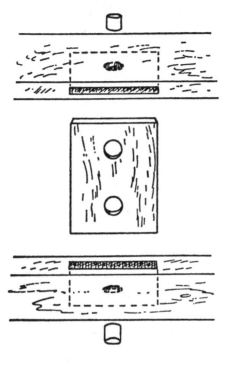

Fig. 49

that Odysseus 'did as much joinery on his improvised boat as a skilled craftsman would put into the hull of a broad merchant ship', he simply means that he did the job thoroughly and without skimping. Three methods were used to keep the tenons in place, and thus hold the edges of the planks snugly together. First, they were shaped to fit tightly into the mortises and were tapered slightly towards each end, so they could therefore be called, both in Greek and in Latin, 'wedges'. Secondly, some kind of waterproof adhesive, made from resin, was used. Finally, to make quite sure that the tenons would not slip out of the mortises, a hole was bored through each plank and at each end of the tenon, and round wooden dowels driven through.

To hold the planks in the appropriate position for joining together, and to give them the correct curvature, a series of 'timber-clamps' (in Greek *dryochoi*) were fixed around the keel during construction, but they were outside the hull, and did not eventually form part of the ship. In fact, the usual practice was to shape and fit the boat-frames inside the hull after it was complete or almost complete—the reverse order of construction from that of a clinker-built boat.

In order to strengthen and protect a hull made in this way when the sea was very rough, the Greeks used devices called *hypozōmata* —'under-belts'. The most famous mention of them comes in the account of St. Paul's voyage (Acts, Chapter 27, v. 17). The generally accepted view (though there has been much argument about it) is that the 'under-belts' were heavy cables, running along the outside of the hull from stem to stern, which could be tightened up during an emergency by means of windlasses. The modern term for this practice is 'frapping' (but see p. 227).

The shape and size of the hull varied greatly, according to the function to be served by the boat, and a wide range of terms for different types of vessel seems to have been used rather carelessly and inconsistently. There were, however, two basic types of hull, for which the simplest distinguishing terms were 'long ship' and 'round ship'. The 'long ship' was essentially a warship or pirate vessel, designed for rowing at high speed in action, though sails could be carried for cruising or long voyages. The 'round ship' was a merchant vessel for sailing only, apart from the smallest ones which were rowed on rivers or in harbours.

The differences in hull shape of the two types were exactly what one would expect. The 'long ship' was slender, with a length-to-beam ratio of about 10:1. A characteristic feature to be seen from the side was the tall sternpost, which curved upwards and forwards. Ancient ships were normally beached stern first, and in many illustrations a short landing-ladder is shown tied to the sternpost. At the bow another even more distinctive feature was a large projecting ram, which looks almost like an extension of the keel and keelson. It could take various shapes. In older vase-paintings it is like a rudimentary figurehead, but in the fifth century it apparently had two chisel-like blades, one above and one below the waterline. Their points were sheathed in bronze to increase their destructive power. The join between the ram and the stempost was shaped to reduce water resistance, so that the whole structure acted both as an armament and as a cutwater.

The 'round ship' was much broader, with a length-to-beam ratio of about 4 : 1. Since it was normally under sail, and in any

case unable to engage in fighting tactics, it had no ram. In order
to increase the cargo capacity the bows and quarters were full and
rounded, and there was usually a broad expanse of bottom which
was almost, if not completely flat. The resistance on such a hull as
it went through the water was obviously greater, but that would
not be much of a problem at the sort of speed they could hope to
achieve.

Yet another distinguishing feature of ancient Greek and Roman
ships was the steering mechanism. Instead of a rudder, hinged on
the sternpost, they used a steering oar at the side of the hull near
the stern or (more often) one on each side. In illustrations throughout the classical period the blades appear rectangular in shape,
and on 'long ships' they usually trail in the water at an angle of
about 45°, projecting a little beyond the stern. If the ship had to
be beached stern first it was necessary either to remove the steering oars altogether before grounding, or at least to hoist the blades
up into a horizontal position well above the waterline.

The method of mounting the steering oars has been the subject
of some argument, and it is difficult to reach firm conclusions
from the existing evidence. On warships, it appears that they had
loops of rope ('strops') attached at the appropriate point, which
were lowered onto some sort of vertical peg on the outside of the
hull when the ship was launched, and shortly before landing, the
oars were lifted off their pivots and shipped. With this arrangement, they could have been operated as in Fig. 50a, and one Greek
expression used for the helmsman's movements is 'pushing out'
and 'pulling in', which might refer to this. There is also, however,
evidence to suggest that on some vessels, particularly large merchantmen, the steering oars were nearly vertical and had their
round shafts mounted in a crude kind of bearing, in which they
were twisted round by means of tillers slotted into their top ends
(Fig. 50b). The tillers may or may not have been joined together
by a connecting rod, so as to turn together. This arrangement
would fit better with the statement in a passage of Lucian, to be
discussed later, that a diminutive and elderly helmsman was able
to control a very large merchantman and steer her into port without any great effort.

One would naturally expect that the warship would require
more powerful steering apparatus than a merchantman, for two
reasons. Firstly, because it was longer and narrower, the thrust

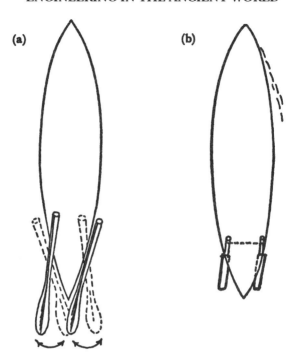

Fig. 50. Methods of steering.

required to turn it around would be greater. Secondly, for the purposes of ramming it would be necessary to turn in a very tight circle on occasion, but if the crew were sufficiently competent, it might have been possible for the rowers on one side to back water and swing her round in that way. For a long time it was believed that steering oars were less efficient and more difficult to control than a modern rudder, but recent experiments with replicas of Viking ships (which had a single steering oar, not unlike the Greek and Roman ones) have shown that this view was quite mistaken. Among other advantages, the larger steering oars on merchant-men turned about a central bar, and were roughly symmetrical, with the result that the water thrust evenly on each half, and in opposite directions, which made the helm lighter.

The characteristic warship of the Greeks and Romans was a 'long ship' (as described above) propelled by oars during naval actions. Mast, yard and a square sail could be used for long sea voyages, but for maximum performance they were normally removed and left ashore when the ships were launched against the enemy.

Apart from two modern refinements, oars and rowing technique have not changed much since ancient times, and a Greek rower from one of the ships which fought at Salamis would find himself much at home in a lifeboat from the early part of this century. One novelty he would notice would be the U-shaped rowlock, which was unknown in antiquity. Greek and Roman oarsmen had fixed vertical pegs ('tholepins') against which the oar was pulled, and to which it was tied by means of a leather strap. The Greek word for 'tholepin' was *kleis,* meaning 'key', and where they are shown in illustrations, they are shaped as were early keys, like an inverted L. Ancient writers on mechanics refer to the tholepin as the fulcrum in their lever analogies. If the oar 'rolled' around a circular peg, without rubbing too much, wear need not have been too severe. There is also a remark in an ancient commentary on a passage in Aristophanes (*Acharnians* 96) that the function of the leather strap was 'to prevent rubbing on the woodwork', meaning (presumably) the gunwale. This is consistent with the fact that it was called *tropos* or *tropoter,* meaning 'turner' or 'twister'. In addition to keeping the oar up off the gunwale, it kept it in position under the horizontal part of the 'key', and also served to prevent it from being lost overboard if the rower (through 'catching a crab' or for other reasons) momentarily lost his grip on it. On the other hand, it was not permanently fixed to the oar, being listed as a separate item in the rower's kit.

Our ancient Greek would be used to rowing from a simple fixed bench, and would be totally unacquainted with the sliding seats of a modern racing boat: to prevent blisters on the bottom, each rower was issued with his own personal cushion.

Illustrations show an almost infinite variety in the numbers and arrangement of rowers in various types of vessel, but in the case of the oared warship it is possible to trace a continuous development in design from 'Homeric' times (seventh century B.C.) until the end of the classical period (late fourth century B.C.), culminating in the best-ever design of ancient warship—the trireme. After that, development took a different line altogether, striving for size and impressiveness rather than performance. Then, in turn, came a reaction against this trend, and a return to earlier, simpler designs, which persisted until the end of Roman sea-power in the Western Mediterranean in the late fourth century A.D.

Homeric rowing ships had single banks of oars arranged symmetrically on either side. Small vessels for rapid transit, conveyance of despatches or important passengers had about twenty rowers, and larger vessels, such as those which conveyed the Greek contingents to Troy, had up to fifty. Such a vessel was called in Greek a *penteconter* ('fifty-oarer'), and at a rough guess, it might have been about 100ft (30.5m) in length overall and about one-tenth of that in the beam.

From about the middle of the eighth century B.C. there appear in vase-paintings ships with two superimposed banks of oars. This change was accompanied by another, which must have been known in Homer's time but is not mentioned by him—perhaps because he wished in this matter to maintain a 'genuine antique touch'. In order to make it possible for ships to carry a landing-party, or to fight at sea by drawing alongside and attacking each others' crews (Homer has no occasion to mention either of these operations), a raised deck was built, running the whole length of the ship but not the full width. A space of 3ft or so (1m) was left undecked along either side to give headroom for the rowers and avoid the danger of their being trapped below deck at the mercy of a boarding-party.

The next developments are not well documented in literary sources or illustrations, and the exact order in which they took place cannot be established with certainty, but from the design which eventually emerged we can infer what the key changes were. When the second bank of rowers was added, it was probably above the first, and since the original bank rowed at gunwale level, the second bank must have been roughly level with the raised deck. Very soon, if not from the start, designers found that it was better to arrange them 'in echelon' (Fig. 51a) than to have one directly above another. The fault of this design, however, is obvious. The crew themselves made up a considerable fraction (about one-fifth) of the total weight of the craft and, being positioned so high up, they would cause it to be top-heavy and liable to capsize.

A better alternative arrangement was to place the second bank below the first. To do this, a row of holes ('oarports') had to be cut in the hull, and a reinforcing bar fixed below them to hold the tholepins and take the thrust of the oars (Fig. 51b). This arrange-

ment lowered the centre of gravity and made the ship more stable, but it caused another problem. If the sea was choppy, or if the
ship heeled over during a sharp turn, water might be shipped
through the ports. To prevent this, a leather bag (in Greek, *askoma*)
was fixed around the oar and (somehow or other) around the
edge of the port, perhaps using the top half of a small animal's
skin, the oar shaft passing through the neck-hole.

Fig. 51. Arrangement of two banks of oars.

Eventually (the exact date is uncertain, but probably some time
in the latter half of the sixth century B.C.) these two designs were
combined to form a new type of vessel which became the standard warship for more than a century. The Latin name for it was
triremis and this, anglicized as 'trireme', is more commonly used
than the Greek form *trieres*. Both of them mean (probably but not
quite certainly) the same thing—a 'triple-rowed' or 'triple-oared'
vessel, that is, one with three banks of oars. There has been a lot

of controversy over the past eighty years or more about the design and rowing arrangements, and it would be tedious and time-consuming to follow the history of the various theories which have been advanced. In this context only the most recent—and best—solution to the problem, first advanced by J. S. Morrison in 1941 and now generally accepted, will be discussed.

It is probable that in the prototype version the lines of rowers in the three banks, viewed from the bow or stern, were vertically above one another—the top bank level with the raised deck, the middle bank at gunwale level, and the lowest bank rowing through ports in the hull. This, however, would mean that the oarsmen of the top bank would either have to use extra long oars, or else hold them at a very steep angle to the water, and manipulate them very much in among those of the lower banks. This problem was solved by building on an outrigger (called in Greek *parexeiresia*—'out-along oarage') which projected beyond the gunwale. The rowers' seats were placed directly or nearly over the gunwale, and the rail which carried their tholepins was about 2ft (60cm) outside the gunwale and perhaps a foot (30cm) or even less, above it.

There were a number of advantages in this arrangement. First, the rowers in the top bank, because they were to one side ('outboard') of those below them, did not have to be so far above them vertically, and this lowered the centre of gravity, making the ship more stable without increasing its beam. Also, it enabled them to use oars of the same length as those of the other banks, without having to hold them at a very steep angle to the water. Even so, their task was considered the hardest, and they were on occasions given higher pay than the others. They were called *thranitai,* or 'stool-rowers'.

Those on the middle bank were called *zygioi,* or 'thwart-rowers', and those on the lowest one *thalamioi* or 'hold-rowers', who had the most unpleasant and dangerous position. If the ship got badly holed, they were the most likely to be drowned or captured by an enemy boarding-party. Also, as Aristophanes points out with homely vulgarity (*Frogs,* 1074), they sat with their faces rather close to the bottoms of the *zygioi* above and in front of them. Their oarports were only about 18in (45cm) above the waterline, and even with efficient oar covers, they must have got quite wet.

No remains of ancient triremes have as yet been found by under-water archaeologists, though there may be a simple reason for this, which is discussed later. But from illustrations, from inscriptional evidence and from the remains of shipyards in the Peiraeus (the port of Athens) it is possible to work out rough measurements for a trireme, as shown in Fig. 52, and also the number and arrangement of the rowers. The last item of evidence has recently been

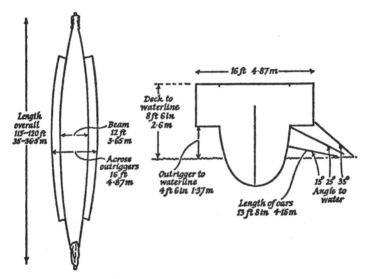

Fig. 52. The Greek trireme.

reinforced by the discovery of similar foundations in ancient Carthage. There were 27 rowers on each side in the lower and middle banks, and 31 in the top bank. Since the hull curved gently upwards at the stern, restricting the space for the lower banks at that end, it is probable that the extra four *thranitai* of the top bank were placed there, and not at the bow, one of them, close to the helmsman, acting as 'stroke oar'. This made a total of 54 + 54 + 62 = 170 rowers, which, added to the helmsman, the boatswain (in Greek *keleustēs,* the 'giver-of-orders'), the ship's commander and junior officers, and a small party of marines, made up a total crew of just over 200.

Looking back from the destroyers and motor torpedo-boats of the twentieth century, it is easy to fall into the belief that ancient warships were absurdly slow and clumsy. This was by no means the case, but the reason why such a belief persists is that, until very recently, no real attempt has been made to assess their performance. To build replicas of a *pentekonter* and a trireme, and to establish figures by experiment, would be prohibitively expensive, but it is possible to make theoretical calculations of their propulsive power and water resistance, from which maximum speed can be estimated. If, moreover, the same calculations come out right when applied to a modern rowing boat of known and accurately measurable performance, we may feel confident that they give a close approximation to the speeds of ancient vessels. A brief outline of the method is given in the Appendix at the end of this chapter.

The estimate for the maximum speed of a *pentekonter* comes out at 9.5 knots—just about the same as a modern racing eight. Specially trained crack crews have reached 10.65 knots in still water, but figures worked out from the Oxford and Cambridge Boat Race are misleading, as that race is deliberately timed for a favourable tide of 3–4 knots. For the trireme, the figure is even higher—about 11.5 knots. The sight of a fleet of these vessels cutting through the water at such a speed must have been most awe-inspiring, and the impact of their rams on enemy ships devastating.

That, however, was the maximum speed which could be attained by a well-trained crew in good physical condition, and could only be kept up for perhaps ten minutes or so. If they had to row over a long distance, they must have reduced their output to about $\frac{1}{10}$ h.p. each, which would cut the speed down to just below 9 knots. This was probably the sort of speed achieved in the famous dash to save the Mitylenians, described by Thucydides (III, 49). The Athenian Council had passed a decree that all the male citizens of that city-state (which had revolted against them) should be put to death and the women and children sold into slavery, and a trireme had been sent to convey the order to the Athenian commander there. The next day saw a change of heart, and a meeting of the Assembly revoked the decision by a small majority. A second trireme was sent out immediately to try to catch up with the first and countermand the order.

Thucydides does not tell us how long the voyage (of about 186 sea miles or 345km) took, but he does say that the first trireme had a start of 'about a day and a night—say 24 hours—and that it had been in harbour at Mitylene just long enough for the despatch to be handed over when the second trireme arrived. The first, he says, 'was not hurrying', and the crew of the second were promised a large sum of money by the Mitylenean envoys and their friends in Athens if they made it in time. They were provided with food and wine, which they took while actually rowing, and (later on perhaps) worked in relays, some sleeping while the rest rowed on. There are two possible ways in which this might have been done. If all superfluous gear was removed, and no personnel carried except (say) one senior officer with the despatch and a few men to prepare and serve the food, a spare crew for one bank of oars could have been carried on board. 54 rowers would make only about 20 'extra bodies' above the normal fighting complement, and would not have slowed the ship down much. All three banks could then have been manned all the time, the men rowing four hours on and two off', or some such arrangement. If so, they could have kept up a speed of 8–9 knots and done the journey in 21–23 hours. By a lucky chance (says Thucydides) the sea was calm and there was no headwind. The first trireme, cruising on one bank of oars and not hurrying, might well have averaged as little as 4 knots, and taken 24 hours longer.

Another possibility is that the normal complement of rowers was carried, and each bank had one or more rest periods, getting what sleep they could in the bilges or on the deck while the other two rowed on. This would reduce the speed to about 6.5 knots, but if they all rowed for the first six hours and the last six, they would still have made it in just over 24 hours. Navigation need not have been a problem. If they set out early in the afternoon (which is quite probable), they could have rounded Sunium and reached the straits between Andros and Euboea by nightfall. From then on they would be on open sea and could have steered by the stars, keeping a sharp lookout for Psara if they went to the North, or Psara and Chios if they sailed between. Sunrise would come between there and the southern tip of Lesbos, and a six-hour stint by all three banks from about 8 a.m. onwards would bring them to Mitylene not long after midday. There were, after all, many hundreds of lives at stake.*

*On the modern map by which these distances were measured (Hallwag, Bern 1972) the trip by island ferry from Athens to Mitylene is marked 'about 20 hours'.

To describe in detail the tactics and methods of Greek and Roman naval warfare is far beyond the scope of this volume, but it may be helpful to list a number of limitations imposed on fighting ships and their commanders by design and technical factors.

First, the trireme was not a 'heart of oak' vessel. It was made—for lightness combined with strength—mostly of soft woods such as pine and fir. These were not plentiful at all times in Greece, and the Athenians imported large amounts mainly from Thrace and Macedonia. One result was that the hulls tended to soak up water, and no effective surface sealing agent seems to have been discovered, unless a substance called *hypaloiphē* ('under-paint'), listed among ships' chandlers' items, was some kind of varnish or sealer. Consequently, all triremes were beached and man-handled out of the water as often as possible. One of the most serious problems for the Athenians in Syracuse harbour (as described by Thucydides VII, 12–15) was that the enemy could launch their ships at any time they chose, whereas the Athenians, having no reserve of vessels, had to keep all of theirs in the water all the time in case of sudden attack. As a result, their hulls had become waterlogged, and they could not make anything like their maximum speed. The shipyards with sloping slipways at Zea in the Peiraeus and at Carthage were partly for shipbuilding, but also for drying out and cleaning the hulls.

Secondly, the trireme was designed for speed at the expense of stability. Its performance in calm waters was very high, but in anything like a rough sea it was neither fast nor safe. This meant that it was rarely possible for warships to maintain a sea blockade over a long period. The chances were that a stormy spell would set in, during which the heavier, slower merchant ships could sail in (provided the wind was in the right quarter), but the risk of launching warships against them would be too high. There were also other reasons, which are mentioned later.

Thirdly, being made as light as possible, they tended when holed by the enemy to settle in the water without actually sinking. The Greek word *kataduein*, which is almost invariably translated as 'sink', in fact means no more than 'dip' or 'lower' and when a Greek writer wishes to indicate that something 'went to the bottom', he generally uses a different word. So, when triremes were holed in a sea battle, though they had become absolutely useless as fighting vessels, the combatants went to great lengths and some risk to recover the wrecks. For example, in the engagements off Naupactus in 429 B.C. (described in Thucydides II, 83–92) some Athenian triremes were driven ashore and damaged—quite seriously, it seems—by the Corinthians, who then tried to tow them away against opposition from the Athenian allies on the shore. Later, when the tables were turned and the Athenians got the upper hand, they recaptured the damaged ships and towed them back to their naval base. This may also account for the fact that no actual remains of ancient triremes have yet been found by underwater archaeologists. Such woodwork as they have found has come from merchant ships, and has been carried to the sea bottom by the weight of the cargo. It could be argued that the heavy bronze casing of the ram would be likely to drag a holed ship under, but this is certainly not the impression one gets from eye-witness accounts. If the ram broke off, it would of course sink rapidly on its own.

From this and from other evidence we can appreciate that another advantage of trireme construction was that it lent itself to extensive repairs and re-fitting. When Polybius describes the loss of a Roman fleet in a storm off Sicily, and adds that the damage was so great that the wrecks were not worth recovering, he obviously regards this as exceptional. This in turn also accounts for the fact that a number of ancient warships—both individual vessels and fleets—seem to have been in service for a remarkably long time. About twenty years was quite normal for a trireme, but we do not know how many re-fits it would undergo in that time, or how much of its original woodwork would have been replaced.

Another very important consideration is that, however well designed the trireme may have been, it was only one half of a partnership, the other being a fit, well trained crew whose morale was high. Ancient historians repeatedly stress that it was difficult to make a ramming attack in narrow waters. The target vessel had to be seen and selected from a distance, and the approach course and speed had to be very finely regulated. The 'window' during which an effective strike could be made was very short indeed, as is shown in Fig. 53. Assuming the modest speed of 9 knots, each ship would travel its own length in about $6\frac{1}{2}$ seconds. If the attacker arrived about 4 seconds too soon (t—4), he could himself be struck and holed by the target vessel, and if he tried to

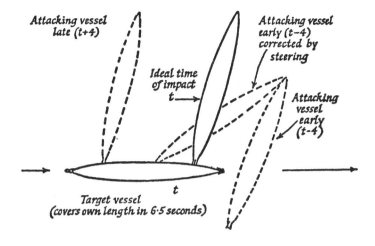

Fig. 53. Ramming.

avoid this by turning to starboard, his ram would strike only a glancing blow, and inflict very little damage. Up to perhaps four seconds after the optimum time (t to $t+4$) some damage, though less than the maximum, could be inflicted. After that, the target vessel was virtually safe, since the speed of impact (being the difference between the speeds of the two vessels) fell off rapidly and the attacker could deliver no more than a mild bump. In fact he would do much better to give up and turn to another target. It is clear, therefore, that during ramming manoeuvres the commander and helmsman would have to rely on the all-out effort and total loyalty of every member of the crew. A mere handful could, if they so wished, upset the rhythm of a whole bank, and make accurate steering and speed control impossible.

The difficulty of finding enough crewmen of the right calibre was a perpetual problem for all ancient navies, and the shortage (more or less permanent) of trained crews made itself felt in various ways. One of the less attractive features of Greek naval warfare was the not uncommon practice of slaughtering oarsmen captured on board enemy vessels or picked up from the water. Later navies, particularly the Roman, tended to rely on bigger, slower vessels, and used boarding techniques in preference to ramming, which meant fewer rowers in proportion to the 'marines'. The bludgeon, one might say, took over from the rapier.

Finally, the trireme was equipped simply and solely for rowing and ramming. There were no officers' cabins, crew's quarters, food store or galley. In fact, more than one naval engagement was won by the commander of an attacking force choosing his time skilfully so as to catch the enemy crews ashore, procuring or eating their main meal of the day. And when large-scale expeditions or troop reinforcements were sent overseas, those who travelled in the warships (and not in the supporting fleet of merchantmen) must have had a very crowded and uncomfortable trip. Also, this made it impossible for triremes to stay at sea for long periods— e.g. to maintain a blockade: they needed a base nearby to which they could return at least once each day.

The history of the oared warship after the trireme is extremely complicated, and beset by lack of evidence and much controversy. It will therefore be summarized very briefly.

The next developments were the 'quadrireme' and 'quinquireme' (in Greek *tetrēres* and *pentēres* respectively). The trireme had three banks of oars, and it might be thought that these vessels had four or five, but there is no evidence for any ancient warship having had more than three. The probable explanation is that a four-rower' had four men in each 'box' or unit of oarsmen, just as the trireme had three—one from each bank, the *thranitai*, the *zygioi* and the *thalamioi* as described above. If so, there must have been more than one man on some of the oars. The most probable arrangement for a quinquireme was two on each oar of the upper and middle banks and one on the lowest. Apart from a short-lived fashion for much bigger vessels, this became the standard warship in succession to the trireme in Hellenistic Greek navies, in the Carthaginian navies of the fourth and third centuries B.C., and in the Roman navies down to the end of the Republic (31 B.C.). By comparison with the trireme it was broader in the beam and had a deeper and a bigger displacement—of the order of 75 tons as against 40 for the trireme, which enabled it to work in rougher weather and to carry more fighting troops on deck—as many as 120 on the Roman ships used in the Second Punic War. It had the additional advantage that a certain number of more or less unskilled rowers could be taken on, by making each of them work alongside a skilled and experienced companion. It was, however, considerably slower and more cumbersome than a trireme.

Our evidence suggests that the quadrireme was introduced early

in the fourth century B.C. and the quinquireme soon after.
About the middle of that century the 'sixer' was developed—
presumably with two rowers on each oar in each bank. Then the
development accelerated rapidly. After the death of Alexander
the Great in 323 B.C. an 'arms race' sprang up between the so-called
'Successors'—his senior generals, who divided his kingdom and
resources among themselves. Almost every denomination—the
'sevener', 'eighter' and so on—up to a 'thirteener' is in evidence
by the end of the fourth century, and by 288 B.C. we hear of a
'fifteener' and a 'sixteener'. The ultimate in these battle-ships came
with a 'twenty-er' and two 'thirty-ers' built by Ptolemy II in the
second quarter of the third century.

In the absence of any real evidence for the disposition of the
rowers in these big ships, scholars have speculated on various
possible arrangements—some of them credible, and others less
so. There is one practical limitation, which applied as much then
as in medieval and Renaissance galleys. A long oar pulled (or pulled
and pushed) by a number of men—known as a 'multiple-rower
sweep'—cannot be more than a certain length. This is simply
because the men near the inner end can only move over a limited
distance backwards and forwards or up and down, and when they
have to stand up, or climb up on a sort of ladder, to get the blade
of the oar into the water, their efficiency is much reduced. In Renaissance galleys eight men to a sweep was found to be the biggest
practical number.

Some of our sources also tell us that the biggest battleships were
'double-prowed' and 'double sterned', and had two helmsmen.
The best (though not the only possible) interpretation of this is
that they were large catamarans with a single solid deck supported
on two hulls. The unsolved question is whether there were, or
were not, rowers on the inward-facing sides and if so, how much
space they had, and how they managed to row effectively.

The drive to produce more and more massive ships had by now
spent itself, and most later Greek fleets consisted mainly of quin-
quiremes, with just a handful of larger ships ('niners' or 'seveners')
used by naval commanders as flagships. There was, however, one
final, futile effort made by Ptolemy IV of Egypt in the last quarter
of the third century B.C.—some 50 years after the heyday of the
big warship. This was a 'forty-er' (in Greek, *tessarakonteres*), a description of which was written up by Callixenus of Rhodes, who

lived in the first half of the next century, and may have seen the vessel still afloat. His remarks are quoted in part by Plutarch (*Demetrius* 43.5) and Athenaeus (V, 203e–204b). Even allowing a lot for exaggeration (and this from natural scepticism, not from any conflicting evidence) the specifications are truly astonishing:

Length overall	425ft	130.5m
Beam overall (or each hull?)	58ft	17.6m
Height from tip of sternpost to water-line	80ft	24.5m
Length of steering oars (4 in all)	45ft 6in	13.8m
Length of oars in top bank (the longest used)	57ft 8in	17.5m

Athenaeus goes on to say that, 'in the course of a test', the following personnel were put on board:

Oarsmen	more than 4,000
Other ratings	400
Marines (on the deck)	2,850
Total	7,250

Plutarch, and a number of modern scholars, have taken these figures to represent the normal crew, but it seems much more probable that they were derived from an experiment designed to see just how many men could be put on board before the ship got dangerously low in the water. It suggests that the total displacement might have been of the order of 1,000 tons.

Given the personnel on deck to man a huge array of catapults and missile-droppers, and to put big boarding-parties on to any captured vessel, such a warship was without doubt unapproachable, unsinkable and altogether invincible. There was just one slight problem—it was also practically immovable. According to Plutarch, it was just a showpiece which stayed at its moorings and never went into action, and although here (as so often) he is complaining about the extravagance of the rich, he may be right.

The evidence on ancient merchant ships is mainly literary and inscriptional, supplemented in recent years by the discovery of a number of wrecks. There are a few ancient illustrations, but this type of evidence is almost completely lacking over the last five centuries B.C., which is in many ways the most interesting period. We are therefore forced to rely heavily on the assumption that ship design was conservative, and that Roman reliefs of the second century A.D. give us a fairly accurate picture of the Hellenistic Greek ships of 400 years earlier, and of those of the classical period before them.

It has already been remarked that almost all ancient cargo ships were under sail. Only the small craft were rowed, such as lighters for bringing in cargo from a big vessel moored off-shore, or for river traffic. There is very little evidence for combinations of sail and oar power, except in warships which appear to be trying desperately to get away from an engagement. There are good technological reasons for this: the requirements of design for an oared vessel conflict on a number of points with those for a sailing vessel, and vice versa. So, if both power sources are used, one or other is being misapplied in the wrong type of ship. The heavy *galeasses* of the Renaissance represent an attempt to compromise between the conflicting requirements, and as such are considered to have been a failure.

The most striking contrast between an early Greek and a modern sailing boat is in the sail itself. Until late antiquity (probably the fourth century A.D.) almost every vessel in the Greek and Roman world, so far as we can tell from illustrations, had a square sail set athwart the hull, unlike the 'fore-and-aft' rig, in line with the keel, which is standard nowadays. A square sail in its crudest form, like a modern spinnaker, is only effective when the wind is blowing from dead astern or nearly so, but even the earliest illustrations show three methods of controlling the area and position of the sail, by which it could be made much more adaptable.

Odysseus' home-made boat, being smallish (he had to launch and sail it single-handed) may have had a fixed mast, but the larger vessel in which he set out on his adventures, like the other Greek ships in the Trojan expedition, had a mast which could be lowered onto a support at the stern ('crutch') when the ship was beached, or in a calm when it was being rowed. On launching, the mast was hoisted up into position and held there by two ropes called *protonoi* ('front-stretchers') running from the mast-head to points near the bow on either side. These ropes combined the functions of a 'forestay' and 'shrouds', preventing the mast from bending backwards or sideways. If they snapped (this occurrence is described in *Odyssey* XII, 409–12) the mast could fall into the stern and strike the helmsman. It was supported against a strong following wind by the 'braces' described below.

The foot of the mast rested in some kind of socket on the keel. When it was hoisted into position ('stepped') and the stays were made fast, the horizontal yard from which the sail hung was hoisted by its central point, the sail having been furled up close to it. We gather from Homer that in early times the ropes used for this purpose ('halyards') were made from plaited ox-hide thongs. It is unlikely that pulleys were used much before the fifth century B.C.—the surviving remains of them are from much later. Before that, smooth metal rings might have served instead. The position of the yard was controlled by two ropes ('braces', in Greek 'top-ropes', *hyperai*) running from its ends down into the stern. It could be set square for a following wind, or swung around either way to a position approaching that of a fore-and-aft rig, its movement being limited by the *protonoi*. It could also be tilted down at one end and up at the other. On large merchantmen in later times it thus served as a crude substitute for a crane for loading and unloading.

To spread or furl the sail, and to control its effective area, a set of light ropes ('brails') was used. These were attached at one end to the foot of the sail at intervals of about 1–1½ft (30–45cm). From there they ran vertically in parallel lines up the front of the sail, being held in position there by rings sewn on to the sail. Then they passed up over the yard and down into the stern area. In a number of pictures one man is shown holding about half of them as a bunch in one hand, so clearly, it did not require much of a pull to furl the sail. It may even have been necessary to hang weights on the foot of the sail to make sure that it spread itself in a very light breeze when the brails were slackened.

The third method of controlling the sail was by means of two strong ropes ('sheets', in Greek 'feet', *podes*) running from the bottom corners of the sail. On a small boat, if the wind was not directly astern, the one on the windward side was usually made fast to a cleat, while the helmsman held the other (the 'lee sheet') in his hand. Pulling this rope taut has the effect of catching the full force of the wind in the sail, and to do so in a squall is to court disaster—hence the remark in Sophocles' play quoted at the start of this chapter.

The Mediterranean and its weather placed severe restrictions on any seafaring, particularly under sail. Having no magnetic compass, Greek and Roman sailors navigated by the stars at night, and by landmarks in daylight, and so, in anything less than 'very good' visibility, they were very much at risk. Storms are more frequent and severe in winter, but keeping to the summer season does not remove that danger altogether. The earliest Greek lore, preserved by Hesiod, recommended that voyages should be made between June 21st and August 10th—a very curtailed season, even allowing for the fact that he was a farmer, who had made only one short voyage (which he had not enjoyed) and had all the pessimism of his profession. More adventurous seafarers regarded late March or early April as 'a bit risky, but possible', from the beginning of June to mid-September as 'safe', from then until early November as 'doubtful' and the rest of the year 'definitely out', except for dire emergencies. The Garrulous Man, number 3 of Theophrastus' *Characters,* among various other platitudes about rising prices and the weather, is wont to say 'Just fancy— the sea was navigable from the Dionysia (late March) onwards!' For most of five months in every year, the entire commercial life of trading ports effectively closed down.

The later developments in the design of merchant ships show two or more auxiliary sails—a small square one on a sort of bowsprit projecting beyond the bow (called *artemon* in Greek) and a triangular one above the mainsail ('topsail', in Latin *siparum)*. But at all times the main propelling force came from the single, big, square mainsail. The performance of a vessel with this type of rig in various wind conditions has been the subject of a good deal of argument, and it would seem that some practical experiments, which could be carried out without great difficulty or expense, might enable us to get a more accurate picture.

It is generally agreed that Greek and Roman sailing vessels, given a good following wind, could make speeds of about 4–5 knots, with 6 knots being exceptional, but not impossible in ideal conditions. It should be remembered, however, that such evidence as we have tells us the average speed over quite a long voyage— of the order of 300 nautical miles or more—and may conceal a wide variation of speed between spurts and lulls. A square-rigged vessel could sail quite effectively, though not so fast, with the wind on either beam (at right-angles to the desired course, Fig. 54a). This was achieved by using the devices described earlier. The yard was braced around aslant to the wind, and the 'sheet' on the windward side was let out so that the wind carried it forward of the mast. The other ('lee') sheet was drawn tighter (a winch would have been used for this on the bigger ships) and used to control the sail in such a way that the main thrust on the mast was forward.

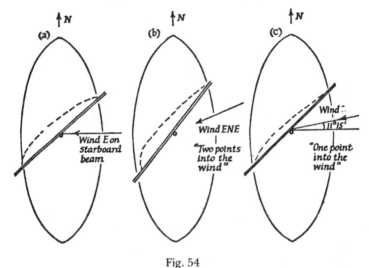

Fig. 54

But however well this was done, two side-effects were inevitable. The vessel would drift sideways through the water, off course to some extent, and would also 'head into the wind'—that is, turn clockwise if the wind was on the starboard beam—and correcting this by the steering oars would not be easy. In fact, the ancient remedy is described in the *Mechanical Problems* attributed to Aristotle. Apparently the brails were used to shorten up the leeward half of the sail (the part astern of the mast) thus reducing the area of sail directly confronting the wind, and enabling the forward half, which tends to turn the ship in the opposite direction, to counteract the 'heading' tendency.

The brails were also used when the wind was extra strong. If it was due astern, the middle of the sail was brailed up, leaving only a small area spread at each end of the yard. If a really sharp squall blew up suddenly, the whole sail could be furled quite quickly, leaving only the mast and yard to catch the wind, and this could be done from the deck, without any of the crew having to 'go aloft', as they had to in later sailing ships. Alternatively, it is sometimes said that the yard itself was lowered part-way down the mast, but this seems a clumsy and slow manoeuvre by comparison.

So far we have dealt with conditions in which the wind was 'following', or on the beam. When it blew from ahead of the beam, things became more difficult. The action described above, of swinging the yard around and adjusting the sheets, could be carried a little further, so that the ship actually sailed slightly into the wind, but how far this could be taken is a matter of doubt and guesswork. The usual approach to the question is to start from the capabilities of square-rigged vessels of the eighteenth and nineteenth centuries. These are known to have sailed, with some difficulty, what is variously called 'six points off the wind' or 'two points into the wind'—that is, at the angle shown in Fig. 54b. On the assumption that Greek and Roman seamen were less competent, it is usually said that they could only manage 'one point into the wind'—that is, with the wind just over 11° forward of a line at right-angles to the keel (Fig. 54c). It would be difficult to challenge this assertion without experimental evidence. What is certain is that they would have tried their utmost to improve the performance of their ships in this respect, for reasons which will become clear.

What happened when the wind was too far ahead for the braces and sheets to cope is very obvious from a number of passages in Greek and Roman authors. The ships resorted, as sailing ships have done ever since, to 'tacking', which involves setting a course alternately to the left and right of the destination (the 'port tack' and 'starboard tack' respectively). If the wind is dead ahead, this means a symmetrical zig-zag course, and it is at once obvious that here is a very slow way of travelling. Even in a fair breeze, a ship sailing close to its maximum angle into wind cannot make its full speed, and it has to cover more than five times the direct linear distance. In theory, geometrically speaking, it makes no difference how many tacks are made—four short tacks from A to B cover the same distance as two long ones, but in practice there are a number of other considerations. The business of changing from one tack to another is hard work, and involves a lot of organization, and the vessel slows down and loses time during the manoeuvre; naturally, therefore, the tendency is to make as few tacks as possible. But the room may be limited in a channel or strait, or if the ship is following a coastline, and the helmsman does not wish to lose sight of land. Again, it is difficult to make changes of tack in darkness or poor visibility. It is also unwise to go out to sea on a long tack if there is a likelihood of the wind changing before the completion of the second tack. All these considerations have to be weighed against each other, and in the ancient world, without even a compass or rudimentary charts, it must have required great skill and long experience to make correct decisions on the best course.

If the wind is not dead ahead, the tacks become asymmetrical and in some conditions the second tack actually takes the ship away from its destination, though this is more than offset by the fact that the first tack is longer.

So far, it has been assumed that one point (11°15') into the wind was the maximum capability of an ancient square-rigged ship. It is clear, however, that even a very slight improvement on that would make a marked difference to the distance travelled and the time taken. For instance, with the wind dead ahead an improvement of just one degree—too small to be detected without quite sophisticated apparatus—would shorten the total tack distance by 8%, and a further improvement of 1° would shorten it by 15% altogether. If the ancient navigators could have managed one-and-a-half points into the wind, the total distance would have been shortened by $\frac{1}{3}$, which would have made a big difference on a long voyage against the wind, when the ship was tacking for most of the time, and many such voyages were regularly made in the ancient world.

For example, from Italy (Brindisi or the Straits of Messina) to Alexandria was a 'downhill run'. The prevailing wind during the sailing season was (and still is) N.W., and, if it blew steadily, merchant ships could make the journey in 18–20 days, at an average speed of just over 2 knots. According to Pliny (*Nat. Hist.* 19, 3–4), a small, fast sailing boat could make it in 9 days, which represents an average of about $4\frac{1}{2}$–5 knots. The return trip, however, was very different. They had to beat into the wind for almost the whole voyage, making it much longer in distance and time— anything between 40 and 65 days, or even more. The longer times probably indicate a number of weatherbound delays in harbour, rather than a very slow rate of sailing, but even so, a very slight improvement in tacking performance would have shortened such voyages by days, or even weeks.

An account of a voyage on precisely that run, made by an unusually large merchantship, is given in Lucian's dialogue *The Ship or Human Wishes*, written in the second century A.D. The opening passage is clearly intended as a parody of the opening of Plato's *Republic*, but the description of the ship and its voyage are introduced entirely for their own intrinsic interest, and the satirical *genre* does not call for any distortion or exaggeration in that part of the work.

The ship was named the 'Isis', and was normally on the grain run from Alexandria to Rome. Its measurements are given as:

length overall	182ft	55.5m
beam 'more than'	45ft	13.9m+
height from deck to bilges	44ft	13.4m

and from these its cargo capacity has been estimated at about 1200 tons. Even so, it was steered into port by 'a little old man, who turned the huge steering oars with a slender wooden rod; he had curly hair, receding at the front, and his name was Heron'. All this sounds (as Lucian intended) very circumstantial.

They had started from Alexandria in a light breeze—apparently a little W of NW, and sailed roughly NNE, sighting *Acamas* (Cape Arnauti, the western tip of Cyprus) on the seventh day. This is about 250 sea-miles, and represents a speed of about $1\frac{3}{4}$ knots—quite reasonable for a heavy vessel on a port tack in a light breeze. Then things went disastrously wrong. A westerly gale blew up. Though they would probably have been making for *Anemourion* ('Windy point', now Anamur on the S. coast of Turkey), they were instead carried 'aslant' (*plagioi*, at right-angles to their intended course) and ended up at Sidon (in the Lebanon, about 20 miles south of Beirut). The change of wind direction need not have caused this change of course, but presumably the gale caught them before they had rounded the tip of Cyprus, and was too strong for them to beat against. The captain (wisely, no doubt) ran eastwards before the gale, and perhaps went further south than he needed to (about ESE), because he could not navigate in the bad visibility. Then, still in rough weather, they went north-wards around Cape St. Andrew and westwards between Cyprus and Turkey (this, to Greek sailors, was *Aulon*, 'The Channel') and got to 'Swallow Islands' (now Gelidonya) 10 days after leaving Sidon.

The coast in that region is very dangerous, with jagged rocks and big breakers. One of the earliest wrecks so far discovered, dating from the Bronze Age, was found there. In addition, they ran one of the gravest risks for ancient seamen—they came on this coast in pitch darkness. But, for a change, they had some good luck. They sighted a fire (perhaps a lighthouse or warning bea-con) which told them they were nearing land, and 'one of the Dioscuri (Castor and Pollux, the patron deities of mariners) set a bright star on the *carchesion* and steered the ship to port when it had been driven close to the cliff'. This is an obvious reference to 'St. Elmo's Fire'—the static-electric brush discharge which appears on the tips of wooden masts and spars in an electric storm. *Carchesion* in the context of a ship normally means 'masthead', but the phenomenon, wherever it may have appeared, was taken as a divine admonition to turn to port (out to sea) rather than stay on course or turn to starboard.

Then, when they had been blown so far off the normal course, the captain decided to give up the trip to Rome, and make for Athens instead. His reasons are not given—perhaps the cargo had begun to deteriorate. Even then, the ship had to 'sail aslant into the Etesian wind'—that is, tack up the Aegean heading against the seasonal Northerlies—and it arrived in the Peiraeus 70 days after leaving Egypt. It should have 'gone to the south of Crete, to the west of Malea (a dangerous area then as now) and should have been in Italy (i.e. Ostia) by that time'.

This was, quite clearly, an exceptionally big merchant ship, though not unique by any means. Hiero II, king of Syracuse from about 265–215 B.C. had a 'super-freighter', named the *Syracusia*, built under the supervision of Archimedes—or so the tradition has it. A detailed description of this huge vessel by Moschion, a near-contemporary historian, has been preserved for us by Athenaeus (V, 206d–209b). It carried a mixed cargo of the order of 16–1800 tons, and also a most formidable array of weaponry and a contin-gent of over 200 marines, since Hiero presumably wished to make it both proof against pirates and capable of blockade-running. In the event, however, it sailed to Alexandria, the only port in which it could be safely berthed, and was there presented to King Ptolemy III, the father of the one who built the biggest-ever warship. So perhaps that rather absurd project was an attempt to 'keep up with the Hieros'.

Ships of this size would clearly present technological problems in building, repair and maintenance. So far as the construction went, it seems that the components of a smaller ship were simply scaled up. In the *Syracusia*, for instance, the pins holding the hull planking to the frames were made of bronze, some weighing $9\frac{1}{2}$lb, and the biggest $14\frac{1}{2}$lb (4.3 and 6.5kg). These corresponded to ordinary nails in a small boat. Similarly, the tenons for edge-jointing the planks of the hull were longer and wider.

Whereas a small boat could be hauled ashore for maintenance, the bigger merchantmen had to stay in the water once they were launched. In fact, then as now, the hull was only part-built before launching, and finished on the water. This raised serious problems of maintenance. The hulls would not only become waterlogged and leaky, but they would also suffer from that scourge of wooden ships, the naval borer (*teredo navalis* as the Romans called it, and zoologists still call it)—the marine equivalent of woodworm or the death watch beetle. Ancient shipwrights avoided using certain woods for the hull, because they were thought to be susceptible to it, larch particularly so, according to Pliny (*Nat. Hist.* 16, 79). But for their bigger ships they adopted a drastic and expensive, but effective remedy. The hull was covered on the outside, first with a layer of linen cloth soaked in pitch, and then with what are called in the ancient sources 'lead tiles'—square or rectangular pieces of thin sheet lead, pinned on the outside. This technique is described by Athenaeus (V, 207b) and the evidence of no less than eight wrecks, ranging from the fourth century B.C. to the first A.D., fully bears out his statement. The cloth layer would improve water-proofing, while the lead layer would improve it still further and, presumably, poison the borers and likewise barnacles or other shellfish which tried to attach themselves. Underwater mainte-nance would thus be reduced to a minimum.

It is a curious fact that no apparent reference is made in ancient sources to 'careening'—the technique of tilting ships over to a steep angle, so that the hull on one side of the keel is accessible for cleaning and repairs. The Greeks and Romans certainly had available the cranes, winches and tackle required for this, but if it was common practice, they seem strangely reticent about it.

There is, however, clear evidence for a much more impressive and sophisticated method, used on Ptolemy's monster warship. Owing to the corruption of a single word in the text of Athenaeus (V 204 c–d) the whole account has always been interpreted as referring to a slipway for construction and launching, despite the fact that it makes complete nonsense as such. What we have here is without doubt a dry dock, crude but workable, for repairing and re-fitting the vessel. A rectangular trench, the same length as the ship and slightly wider, was dug close to the harbour at Alexandria. A foundation layer of squared stone was put down on the bottom at a depth of just over 7ft 6in (3.5m). On top of that, wooden cradles were laid crosswise from side to side, leaving a clearance of just over 6ft (1.85m) above them. Then a channel was dug through to the harbour, the trench filled up with water, and the ship was towed in 'by locally recruited casual labour'. Then the channel was blocked up again, and the water pumped out of the trench 'by means of *organa*', the word regularly used for the forcepump. Thus the ship 'settled safely and firmly onto the cradles'. If it was in fact a catamaran, the whole operation would be much easier and safer. As far as logistics are concerned, such a dock, with the ship inside, might have contained something like $\frac{3}{4}$ million gallons of water ($3.4 \times 10^6 l$) and the pumping-out operation would have required some 500–600 man-hours.

Evidence for the cargo capacity of ordinary Greek and Roman merchantmen comes from various sources, and requires careful interpretation. For instance, a regulation forbidding Roman senators to engage in trade included a clause to the effect that anyone operating not more than two ships of about 12–15 tons burden each was 'not engaged in trade for the purposes of the Act'. This has been taken to imply that vessels of about that size were regularly used, but in fact the exemption level was probably put very low indeed, to discourage attempts at evasion of the law. From certain harbour regulations which have survived, it appears that ships of less than 70–80 tons burden were not allowed to clutter up the wharves in those ports or use the facilities, and when the emperor Claudius tried to improve the Roman corn supply by offering favourable interest rates for shippers and what amounted to free state insurance against loss by storm, the lower size limit was placed at about 65–70 tons burden per vessel. In other words, vessels smaller than that were not considered to be making a significant contribution to the transportation problem, and no wonder, as we shall see shortly.

An ordinary, smallish merchantman seems to have had a cargo capacity of 120–150 tons, and at a very rough guess, it might have been about 60ft long (18.3m) at the waterline and about 20–25 ft (6–7.6m) in the beam. Ships of 400–500 tons burden were by no means uncommon, and we do not know how many very large ones were in service. It is significant that those singled out for special mention are all well over 1,000 tons.

Cargoes were stacked on ships in various ways, as evidenced by underwater archaeology. Grain was usually in sacks, each containing something like a bushel, and liquids were mostly in *amphorae,* the characteristic tall, necked jars from which wrecks of Greek and Roman ships are often first identified. Heavy materials such as stone or metal were stowed on a floor just above the bilges, and they, and the *amphorae,* were packed in place with brushwood. From some wrecks it appears that there were three or four floors, each with its own layer of *amphorae,* one above the other. Some grain ships had bins in the hold, into which the grain was shot in bulk, which would save time on loading, but would add a good deal to the unloading time, as the grain would probably have to be put in sacks to be carried ashore.

Of the countless journeys made by a handful of ships from port to port, carrying anything from marble for building, metal ingots, oil, wine or grain to looted art treasures, we have very little written record. Contracts were often drawn up between the merchant, the ship owner and, in some cases, the banker who put up the money for a trading voyage (and, through a clause cancelling the debt if the ships were lost at sea, provided the ancient equivalent of shipping insurance). But such documents would be destroyed when the transactions were complete. Our main evidence consists of references in speeches (written for private lawsuits) to the usual terms and conditions. One of these speeches, preserved because it was believed to be by Demosthenes himself (*Against Zenothemis,* No. 32) tells of a curiously modern-sounding swindle which involved faking an 'accident' to the ship, but which misfired.

But there was one unique and quite unparalleled transport operation carried out during the early centuries of the Roman

Empire which was on such a huge scale (by the standards of the time) that some account of it survives. Until about the middle of the first century B.C. Rome was able to supplement her home-produced corn supplies from the Western parts of the empire—from Sicily, France, Spain and North Africa. But with the annexation of Egypt after the Battle of Actium (31 B.C.) a new and much greater source was secured (very carefully, as it was under the personal control of the Emperor) for the growing urban population. According to the Jewish historian Josephus (*Bell. Jud.* II, 386) it supplied one-third of Rome's annual needs. The grain was grown in the Nile valley and the upper Delta (mostly in areas irrigated by the seasonal Nile floods), taken down river to Alexandria in hundreds of small boats, and then stored in grain-silos or loaded on a fleet of grain ships. On a conservative estimate, the total quantity exported in normal years was in the order of 130–150,000 tons.

Owing to the restricted sailing season it was impossible for the grain ships to make two complete round trips from Alexandria to Ostia. If about half the fleet wintered in Ostia, they could sail (in ballast, or with light cargoes) at the very start of the season—early April—and get to Alexandria by early May. There they would load up, and sail to Ostia in (say) 65–70 days, arriving before the end of July. If unloaded promptly (this was not always managed) they could leave again late in August, and reach Alexandria before the end of the season. Those which had wintered there would sail at the earliest possible opportunity, and arrive fully laden in Ostia perhaps early in June. If the turn-around was quick enough, they could leave by the end of the month, get to Alexandria by late July, and re-load. But then it might be touch-and-go whether they got back to Ostia before the bad weather set in.

Since the ships required docking and repair facilities during the winter, it would seem reasonable to keep about half the fleet in Ostia and half in Alexandria, but because the second trip to Ostia was not always possible, there would be a tendency for more than half of them to end the season in Alexandria. However, since that meant a greater number available for the double trip the next year, it would not present a problem. The total shipping requirement would be about two-thirds of the tonnage to be carried—say 90,000–100,000 tons. How this tonnage was made up, we have no idea. A possible fleet might consist of about twelve big ships of

1,000 tons burden, 160 of half that size, and 32 of 250 tons. That would be the absolute minimum, leaving no margin at all for delay or loss at sea, or delays in cargo handling.

The first grain ships tacking up the Campanian coast in the spring must have made a brave sight, and Virgil, land-lubber though he may have been, gives a graphic description of Aeneas' fleet sailing across the same stretch of water (*Aeneid* V, 827–32):

> *And now a sweet joy steals across Aeneas' mind;*
> *'Step the masts' he orders, 'spread the canvas aloft!'*
> *All set their sheets alike, and braced the yards*
> *Swinging and slanting now to port, now to starboard,*
> *Brailing up the sail to leeward with each change of tack,*
> *Scurrying along in the fair breeze. . . .*

Appendix to Chapter 6

METHODS OF ESTIMATING THE MAXIMUM SPEEDS OF OARED VESSELS

The propulsive force of a *pentekonter* and trireme can be estimated on the basis of various assumptions. One is that Greek rowers had the same sort of bodily strength as an ordinary fit man (not a trained athlete) today. This enables him to generate about $\frac{1}{3}$ h.p. for a short period (ten minutes or so) and about $\frac{1}{10}$ h.p. more or less indefinitely. Some of this energy is lost because oars, though better than propellers, cannot be made more than about 75% efficient even with the aid of modern science. If we assume that ancient oars were about 70% efficient, this will probably be an error on the low side. The maximum propulsive power for a *pentekonter* would then have been about 11.7h.p. (about 6,500ft lb/sec or 8,700 watts) and for a trireme about 40h.p. (22,000ft lb/sec or 3×10^4 watts).

From these figures it is possible to calculate the total thrust on the vessel (applied at the tholepins) at various speeds. Since the power is the product of thrust × velocity, the thrust diminishes as the speed increases, so the graph of thrust against velocity is a hyperbola (Fig. 55). Also, as the speed increases the resistance encountered by the ship increases, so it will gather speed until the point is reached at which the falling thrust and the rising resistance reach the same level, and exactly cancel one another out—that is, where the two graphs cross in Fig. 55.

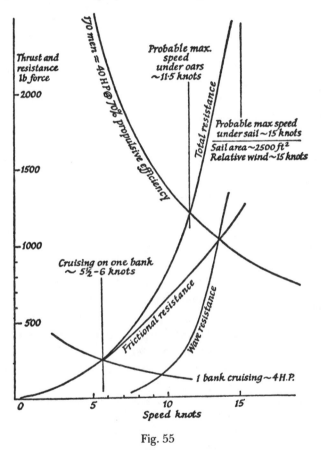

Fig. 55

The water resistance is made up of two components—skin friction and wave-making resistance. The first is, in effect, caused by the water 'rubbing across' the surface of the hull where it is in contact. To reduce it to a minimum it is very important to ensure that the hull surface is smooth, and even a very slight roughness (a grazed patch on the hull or the odd barnacle here and there) can have dramatic effects. It is also important to keep the 'wetted surface' as small as possible. On a *pentekonter* it was about 620sq ft (57.5m²) and on a trireme about 1000sq ft (92.8m²).

Wave-making resistance, which occurs when the boat throws up a wash, can, like the skin friction, be regarded as a reverse thrust on the hull which opposes that of the rowers, but the two thrusts are related in different ways to the speed of the vessel. Skin friction resistance rises slowly in a steady curve (Fig. 55) roughly, but not exactly, as the square of the velocity. Wave-making resistance is almost negligible up to a certain speed, rises in a sharper curve above that, and finally shoots up to an enormous amount. The point at which these changes of regime occur are determined by the speed of the vessel in relation to its length; this factor, the 'relative speed' or 'Froude's number' is given the symbol © and can be worked out very simply from the formula $\sqrt{\dfrac{V}{L}}$, where V is the speed in knots and L the length in feet (at the waterline). When © is below 0.7 wave-making resistance is very slight and can be neglected. Above that it rises in a steep curve, and when © reaches about 1.3 it becomes very large. For the *pentekonter* and the trireme these values of © would represent:

		©=0.7	©=1	©=1.3
Pentekonter	L = 85ft	6.5 knots	9.2 knots	12 knots
Trireme	L = 100ft	7 knots	10 knots	13 knots

Another factor has to be taken into account here—the 'fullness' of the hull. This is calculated from a sort of Platonic Idea of a 'full' hull, which is a rectangular solid whose dimensions are L × B (waterline length × beam) × draught. The fullness of a real hull is its volume displacement divided into that of the solid. For a long, slender ship such as a racing eight this comes out at a very small figure, but for a short, tubby barge it is much higher. Fullness has not much effect on the critical speed at which wave-making resistance begins to rise significantly, but it does affect the steepness of the rise beyond that point. In practical terms this means that Greek warships could cruise at speeds below 6 knots with almost no loss of energy through wave-making, and in fact, the trireme could be rowed at such speeds with only one bank of oars manned. Their maximum speed would be somewhere between 6 and 12–13 knots, and the fact that each comes out near the top end of the bracket is a tribute to the success of the design. Finally, it was totally impractical for anyone in the ancient world to try to increase that maximum beyond 12–13 knots by oar-power. A graphic confirmation of this can be provided by comparing a *pentekonter* and a World War II motor torpedo-boat ('M.T.B.'). These two vessels had about the same length and about the same displacement (20 tons or so). The M.T.B. had twin diesel engines delivering 1,500h.p. altogether, but this enormous reserve of power—more than 90 times that available in a *pentekonter,* enabled it to achieve only $4\frac{1}{2}$ times the speed (42 knots). In that case © is about five, and the energy wasted in wave-making is very high indeed.

Table 3. Greek and Roman measures of capacity.

LIQUIDS

	Greek	Roman		
	1 kyathos	cyathus	0.08 pt	0.045 l
12 kyathoi	= 1 xestes	sextarius	0.96 pt	0.545 l
6 { xestai / sextarii	= 1 chous		5.76 pt	3.27 l
		congius		
8 congii	=	{ 1 Roman amphora or quadrantal	5 gall 6.08 pt	26.17 l
12 choes	= { 1 Greek amphora / amphoreus metretes / or keramion		8 gall 5.12 pt	39.25 l

A Roman amphora jar weighed about $37\frac{1}{2}$–$39\frac{1}{2}$ lb (17–18 kg) when empty. Filled with wine ($5\frac{3}{4}$ gall or 26.17 l) it would weigh about 95 lb (43 kg). Thus 24 Roman amphorae, or 16 Greek, would weigh about a ton.

SOLIDS

	Greek	Roman		
	1 kyathos	cyathus	0.08 pt	0.045 l
6 kyathoi	= { 1 kotyle / or hemina	—	0.48 pt	0.27 l
2 kotylai	= 1 xestes	sextarius	0.96 pt	0.545 l
16 sextarii	= —	1 modius	1 gall 7.36 pt	8.73 l
96 xestai	= 1 medimnos	—	11 gall 4.16 pt	52.36 l
1 modius	= just under $\frac{1}{4}$ bushel			
1 medimnos	= rather less than $1\frac{1}{2}$ bushels			

Weights of wheat, according to Pliny (*Nat. Hist.* XVIII, 66)—i.e. grains alone, after threshing.

Light ⎫   ⎧ 20 librae = 14.4 lb = 6.55 kg.
Medium ⎬ per modius ⎨ $20\frac{5}{8}$ librae = 15.03 lb = 6.82 kg.
Heavy ⎭   ⎩ $21\frac{3}{4}$ librae = 15.7 lb = 7.12 kg.

On average grain, therefore, one ton = about 150 modii or 25 medimnoi.

Land transport

THE shortness of this chapter by comparison with the preceding one reflects—crudely but quite accurately—the unimportance of land transport by comparison with sea transport in the classical world. Any large community which was not self-supporting in foodstuffs grown locally, or raw materials produced locally, depended on importation, which invariably meant importation by sea; two outstanding examples—Athens during the Peloponnesian War and Rome under the early empire—have already been quoted. No such transport operation could possibly have been undertaken by land; even the transit from Ostia to Rome of the grain imported from Egypt, Libya, Spain and elsewhere, was mainly by barge up the Tiber, in preference to road vehicles, despite the fact that this involved re-loading the grain from the granaries on to barges, and rowing or towing them up the Tiber against the current.

Not only was land transport on a much smaller scale in classical antiquity, but its methods and power sources also differed very markedly from those of Western Europe before the introduction of the internal combustion engine. The horse as a traction animal, the mainstay of all land transport in medieval and later times, played an insignificant part in Greek and Roman transport. Its place was taken, for light transport by the mule, and for heavy transport by the ox. Indeed, the whole part played by wheeled vehicles was much less, the most common methods of transport being the human porter, or the donkey or mule with panniers.

There are a number of very logical reasons for these preferences. The human porter (in Latin *saccarius*—'sack man') is much more adaptable in every way than a vehicle or pack animal. He can make access for himself to the hold of a ship, can climb ladders, go along narrow paths or gangways and is, so to speak, self-loading and self-unloading. A very telling piece of evidence came to light during the excavations at Ostia; buildings identified as granaries had access paths and doorways which were almost certainly not designed for vehicles of any kind—they are too narrow, and have sharp bends in them. All the grain, therefore, was taken into and out of these buildings by man-power.

The limitations of human porterage are obvious. The maximum load, in conditions in which it had to be carried more than a short distance (say 40–50 yards), was of the order of 50–60lb (23–27kg). In the granaries and docks at Ostia this would mean a sack containing at the most, about 4 *modii* (see p. 169) or, in Greek terms, $\frac{2}{3}$ of a *medimnos*. Such a load could be carried over distances up to 3–400 yards (275– 365m.) For anything over that, it would probably have been more practical to transfer the load to pack animals for as much of the journey as possible, and unload them by hand at the other end. The rate of movement would be quite slow, perhaps no more than 3 mph (just under 5kph). There was also, of course, the overriding limitation imposed by the available supply of man-power, which became quite short during some periods of the Roman empire.

For larger loads or greater distances the ideal mode of transport is the mule or donkey with panniers. This was the method used by all the small-scale transport contractors in the Greek and Roman world, who were in fact called 'mule-drivers' (*muliones* in Latin, *onélatai* in Greek).

There were several reasons for preferring mules to horses. Mules (and this normally meant the offspring of a mare and a male donkey, not vice versa) are more amenable to the task of carrying a load, being less temperamental than horses, and more easy to train for that type of work. The proverbial stubbornness, combined with the proverbial patience, are much easier to deal with than the high spirits or viciousness of a wayward horse. The Roman soldiers recognized this when they called themselves 'Marius' mules', having been called upon, after Marius' reforms of 101 B.C., to carry with patience a much increased weight of equipment on route marches.* The mule's skin is harder and tougher than a horse's, and hence less liable to damage by rubbing or chafing. It can stand extremes of heat and cold better than a horse, and can survive on less water for a longer period. Its hooves are harder

*Frontinus, *Strategemata* IV, 7.

(equines were not normally shod), and it is much more sure-footed on rocky paths or on the edges of steep slopes. It travels slowly but steadily, at something over 3 mph, and its slowness is partly compensated by the fact that it needs only a short period of sleep—some 4–5 hours in every 24. Thus, if lightly loaded and travelling over fairly easy ground, it can cover something like 50 miles (80km) in a day. By a curious but typical perversity, it tends to walk slowly down a slope and faster up it.

Pack-mules were used extensively until the early part of the present century. From illustrations it would appear that those used by the Greeks and Romans were comparable in size and physique. It is therefore possible to make a rough assessment of their performance.

The mules used by the British Army up to and during World War I were in general between 52 and 60in high at the withers (1.32–1.52m or, as it was usually expressed, '13 to 15 hands'). Extra large ones were as high as 64in (1.62m). In weight they ranged from about 600–900lb (270–410kg). They were reckoned to be able to carry about 30% of their own weight as load, i.e. about 200lb (90kg) for a small animal or 270lb (122kg) for a large one. Naturally, if they were required to cross rough or hilly terrain, these loads had to be reduced accordingly—perhaps by about 25%.

There were further limitations. The load had to be of such a shape and size as to go into a pannier or pouch suspended across the mule's back. It also had to be balanced, which meant that it had to be divisible into two roughly equal portions, one to be carried on each side. For instance, even a very big mule could not carry a block of stone weighing 270lb (122kg); the most it could manage 'in one lump' would be about half that weight, counterbalanced on the other side by a similar weight.

The same considerations apply to donkeys, but the figures for loads and body size have all to be scaled down. Their height ranges from about 3–5ft at the withers (0.9–1.5m, or 'nine to fifteen hands'). One again, ancient illustrations suggest that the donkeys used by the Greeks and Romans were about the same size, or perhaps rather smaller. Thus a small donkey could carry something in the region of 120lb (54kg) in its panniers, and a large one could manage the same sort of load as a mule.

It has already been remarked that a typical small-scale transport contractor in the ancient world would maintain a troop of mules or donkeys, and take on transport jobs from farmers, merchants or anyone else who needed his services. Since his troop could be kept employed for a lot of the time (or so he would hope), it would be worth his while to maintain them permanently, growing some forage crops if he owned any land, or renting grazing land, or even buying fodder from local farmers. His animals would normally be purchased from a mule stud farm. However, if a private individual wished to carry some merchandise of his own (to market, for instance) but was not in the habit of doing so regularly, it would obviously be uneconomical to maintain his own donkeys or mules permanently. In such cases it was common practice to buy a donkey, convey the goods to their destination and then sell the donkey, either along with the goods or else to a dealer in pack animals, if there was one available. This is probably the significance of the phrase in Aristophanes' *Wasps* (367)—'to sell the donkey, panniers and all'.

The wheeled vehicles of the Greeks and Romans fall into two main categories, the heavy farm wagons, normally drawn by oxen, and the lighter vehicles, mainly for passenger transport, drawn by mules, or occasionally by horses. Speaking in general terms, the Greeks and Romans do not seem to have made any very important advances in the design of vehicles. By contrast, the evidence from North-West Europe, in the form of some relief illustrations from France and Germany, and some archaeological evidence from still further north, suggests that Celtic wagon-makers of the early centuries A.D. developed their designs to a highly sophisticated level. It is difficult to find any convincing reasons why this should have been so.

Almost every type of wagon in the classical world was originally designed to be drawn by two animals, and, since the method of attaching them to the vehicle was by yoke and pole, it would seem that the ox-drawn vehicle came well before the horse-drawn. The conformation of the ox, with its 'minor hump' at the withers, suits this method very well, since the thrust, whether it is derived mainly from the front or the back legs, can be taken from that point. All that the harness has to do is to keep the yoke in position. Accordingly it normally took the form of a 'throat-and-girth' harness, i.e. a strap passing around the body just behind the front legs, which served to hold the yoke down and prevent it from riding up over the 'hump', and another passing around the base of the animal's neck, which helped to keep the yoke from sliding backwards, and took some of the thrust (though not much) away from the withers (Fig. 56). The height at which the other end of the pole was attached to the vehicle was such that the animals tended to push slightly upwards on the yoke rather than horizontally forwards. The same was true, of course, when they were harnessed in this way to a plough.

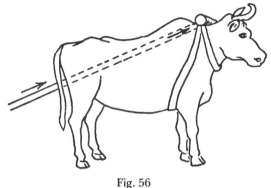

Fig. 56

It is highly significant that this was also the normal method of harnessing equines to vehicles. Here, however, the animal's conformation makes it highly unsuitable. The yoke, having no 'hump' to rest on, can slide back and forth over the withers, and the harness is, therefore, the actual 'power take-off' instead of the yoke. This has all sorts of unfortunate consequences. Because a horse's neck is longer and curves upwards more, the throat-harness tends to ride upwards and forwards; and since it is taking almost all the thrust, it tends to compress both the windpipe, thus impairing the breathing, and the surface blood-vessels, which may interfere with the blood circulation to the brain (Fig. 57.) One attempted solution of this problem was to join the throat-strap and girth-strap together by another strap ('martingale') between the legs. It was eventually solved some centuries later (probably about the ninth-tenth century A.D.) by making two fundamental changes. First, the yoke and pole were replaced by shafts, which ran beside the animal and lower down than the yoke, so that the point of attachment was lower. Secondly, the flexible throat-harness was replaced by the stiff collar, which did not ride up onto the neck, but put the pressure on the shoulders and chest of the animal. At the same time, the shafts ensured that the pull was equalized on both sides.

We have, therefore, the rather strange fact that the Greeks and Romans, though in general intelligent and technologically competent people, used a method of harnessing their horses and mules to vehicles which was clumsy and inefficient, because it had been designed in the first place for harnessing oxen. Why should this be so?

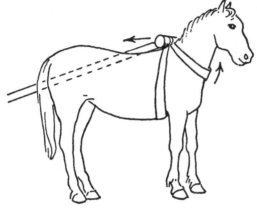

Fig. 57

Here is perhaps an opportune moment to look at some very wide-ranging considerations about the whole history of science and technology.

It has always been the case, and in all probability will always remain the case, that applied science and technology are permanently concerned with catching up with what the society in question sees and recognizes as its immediate requirements. These requirements may be essential (such as improvements in food production) or they may be luxuries (such as power-driven tooth brushes), but in every case they represent something which, at the time, has not yet been made practicable, and it is the task of the technologist to make it so. In the last thirty years we have seen perhaps the most impressive and telling example of this principle in operation. In 1946 it was not possible to build a rocket which would travel to outer space. But two major world powers decided, for various and complicated reasons, that there was a genuine need for them to develop such rockets. Accordingly, the financial resources were made available, and the technologists improved their designs and increased their capabilities until the demands of the two societies were met. This is how technology operates, and this is what it does.

If, therefore, we encounter in an ancient society a situation in which there seems to have been a failure on the part of technology to meet what appears to us an obvious demand, we must ask the following question. Was this because the technologists were incompetent, or was it because the demand itself never really existed? And if we ask this question in the context of Greek and Roman harnessing, what should the answer be?

There is no simple or straightforward answer, but certain facts must be put forward which are crucial to the question. The harness described above was designed for oxen, to enable them to pull a heavy load slowly, almost certainly a plough in the first place. Most of the donkeys and mules of the ancient world would not have been strong enough to pull comparable loads,* and, though there are difficulties in interpreting the evidence, it can be said with some certainty that their horses could not either. There is no evidence whatsoever that the Greeks and Romans had anything comparable to the modern 'heavy horses'—the Clydesdale, Suffolk Punch or Shire. As for the actual size of their horses, there is a curious conflict in the evidence. In black-figure vase-paintings of the sixth century B.C. they are drawn large, comparable in height almost with a modern race-horse, with very slender legs.† In the famous Parthenon frieze, however, the horses in the procession are quite small. It is very difficult to decide whether this is artistic licence—a distortion of scale to accommodate standing men, horses and charioteers within the height of the frieze—or whether it represents the horses accurately and life-size. However this may be, the illustrations of working horses (which are in fact quite rare) do not suggest that they were much bigger, if at all, than mules. The evidence of Roman reliefs suggests that their horses were not much bigger either.

*It is therefore very puzzling to find that Sophocles, in his famous Ode on the ingenuity of man (*Antigone* 338–41) speaks of ploughing the land 'with the offspring of horses'—usually taken to mean mules.

†e.g. Exekias' picture of Castor and Polydeuces (Arias–Hirmer–Shefton, *History of Greek Vase-Painting*, plate 63).

The other important fact is that, where horses or mules are shown in Greek and Roman illustrations harnessed to a vehicle by the system described above, the vehicle is in most cases quite small, and not heavily loaded. There are a few exceptions,* but in general the big horses are drawing chariots (for racing or warfare) and the mules are drawing small wagons. In fact, they would be working very well within their load-pulling capacity, i.e. the load they would be capable of pulling if efficiently harnessed. In those circumstances, the problems of the throat-and-girth harness would not arise, or at least not in an acute form. Four highly-bred horses might in theory have been able to pull a load of 2–3 tons at about 4–5 mph. What they did in practice was to pull a light chariot and its driver—perhaps 4cwt (200kg) altogether—very much faster, which, as we have already seen, was precisely the reason for choosing horses. The typical Roman vehicle for fast travel (carrying one or two passengers) was the *cisium*. Cicero, in his earliest surviving speech, delivered in 80 B.C. *(Pro Roscio Amerino* Chapter 7) speaks of a man making a night-time dash from Rome to Ameria (about 52 miles, 85 km) in 'ten nocturnal hours'. This (to skip the details) would mean an average speed of about 7mph (11 kph). In darkness, and on a 'minor road', this was quite good going. Cicero uses the plural *(cisiis),* which is usually taken to imply a relay of vehicles and horses.

The reasons for using oxen to draw heavy loads are that they are moderately docile (by comparison with bulls, the males which have not been castrated) and sure-footed, and can exert a very strong forward thrust on a yoke—of the order of $1\frac{1}{2}$ times their own body weight. They are, of course, very slow indeed. Under heavy loading, they can travel no more than about 1 mile in an hour, and if there are obstacles in the way it may take them a whole day to cover a mere 5–6 miles (8– 9.5 km). They have, however, a great advantage over horses and mules in the matter of feeding

Bovines are the best adapted of all herbivorous animals. Their intake of food passes first into the rumen, where micro-organisms work on it, and convert almost every type of protein present in the food into microbial protein, which can subsequently be absorbed into the system via the stomach and intestines and fully utilized. Hence the colourful, but accurate description of a cow as 'a giant

*e.g. the four-wheeled cart, laden with sacks, In the Roman relief illustrated in Rostovzeff, SEHRE[2] pl. XLVI, 3.

fermentation vat'. Equines, by contrast, have no rumen, and can only assimilate certain types of protein directly. Therefore, if a horse and an ox of the same body weight consume the same quantity of the same food, the ox is likely to gain quite a lot more nutrition from it. Some micro-organisms are produced in the large intestine of equines, and they act on the (as yet) unused protein (e.g. grass fibres), but after conversion, only a small amount of the microbial protein is then absorbed through the intestinal wall, the greater part being lost with the faeces. In fact, because the animal has to use up its protein supply in order to manufacture the micro-organisms, there is actually a net loss of protein over the whole process: an interesting contradiction (apparently) of Aristotle's dictum that 'nature does nothing without a purpose'.

This is the chemical advantage of the bovine over the equine. There is another physical advantage, that the bovine can ingest about $1\frac{1}{2}$ times as much food as an equine. This works out at about 3% of its own body weight per day in dry weight equivalent, as compared with 2% for a horse. For this reason, a horse cannot survive on a diet of wheat straw alone. If it eats as much as it can possibly manage every day, the nutrition it can extract from that quantity is not enough to keep it alive. An ox, by contrast, can eat $1\frac{1}{2}$ times the quantity, and extract or convert a higher proportion of the protein content. This, however, represents the maximum quantity, assuming that the fodder is freely available and the animal eats continuously *ad lib*. Normally, if the diet is suitably blended, a working animal can derive all it needs from about half that quantity.

The Romans were well aware of these advantages (though not of the scientific reasons for them) and found means of keeping their working oxen on farms without having to feed them anything expensive, or anything which could not be produced on the farm itself. Cato recommends (*De Agric.* Chapter 54) that during the season of their hardest work (the spring ploughing, in March–April) oxen should be fed 15lb (6.8kg) of hay and 15–20lb (6.8–9kg) of 'mash', i.e. chaff, wine-press refuse, etc. per day. When the busy season ends, they can be fed on lupines, vetch and other legumes, which did double duty both as forage crops and as rotation crops to fix the nitrogen in the soil. Later in the summer they were fed on leaves (elm, ash, poplar, etc.) and when that supply gave out, on hay and chaff. Cato was not given to over-generosity in the matter of feeding animals or slaves, but this diet was not inadequate, even judged þy modern standards.

Of the wheeled vehicles little need be said apart from some general points. The wheels were of three types, solid, made from planks criss-crossed, the crossbar type, and the spoked type (Fig. 58). Exactly as one would expect, the solid type ('drums', *tympana*)

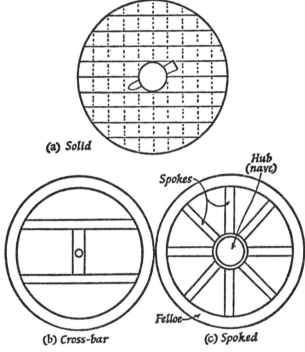

(a) *Solid*

Spokes

Hub (nave)

Felloe

(b) *Cross-bar*   (c) *Spoked*

Fig. 58

were used for the heaviest vehicles and the spoked type for lighter ones, with as few as four spokes on early Greek racing chariots. The crossbar type appears rather rarely. Both two-wheeled and four-wheeled vehicles are well illustrated. The two-wheeled heavy vehicles are usually drawn by oxen for obvious reasons. The load might be exactly balanced over the axle, but the chances are that it would tend to tip one way or the other, and oxen would be better able to cope with either a weight pushing down on the yoke or an upward pull on the girth-strap.

One very much debated question is whether Roman four-wheeled vehicles had fixed or movable front wheels. Illustrations generally suggest that they had a fairly short wheel-base, and could therefore be dragged around corners without much difficulty. There is no evidence to establish the use of a 'swivel-table' on which the front axle is mounted and which can be turned around with the pole (or, on later vehicles, the shafts). It has been pointed out that in most illustrations the wheels appear too close to the body of the vehicle to allow room for them to turn. On the other hand, since most of the illustrations are in low relief, it might be argued that the artist was prevented by his medium from indicating a gap between wheel and body. There is also one scrap of inscriptional evidence in the so-called 'price edict' of Diocletian, the text of a decree fixing maximum prices for a very wide range of goods, issued in 301 A.D. Among parts of wagons there appears the item *columella*, a 'little pillar'. This is the word used for the short spigot on which the rotor of an olive-press turned, and, as it happens, every other use of the word relates to a vertical spigot or post. It would therefore be most natural to assume that in this context it refers to the stub or spigot on which a 'swivel-table' turned.

Another minor question is how the wheels were mounted on the axles, since it is impossible to tell this from illustrations with any certainty. On lighter vehicles it is probable that the axle was fixed, and the wheels turned on a short 'stub' at each end, being prevented from coming off by a 'lynch-pin' passing through the stub (Fig. 59a). To avoid wear on the pin and the wheel-hub, there was normally a metal washer between them. On heavier vehicles it is more probable that the wheels were fixed to the axle, which turned round in some sort of bearing on the under-side of the chassis (Fig. 59b). Both these methods of mounting are in use today—the fixed stub on the non-drive axle of a car, and the rotating axle on railway trucks. The important difference is that wear shows itself in different ways. As soon as the wheel gets slightly loose on the fixed stub, it begins to tilt one way or the other, which means that the area of contact is reduced, and the wear is concentrated at one end of the hub at the top, and the opposite end below (Fig. 60). With the revolving axle, however, the wear takes the form of a groove in the axle where it rubs on the bearings, but does not cause tilting of the wheels, and is therefore safer if the vehicle

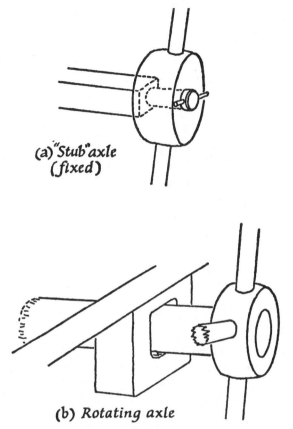

(a) "Stub" axle
(fixed)

(b) Rotating axle

Fig. 59

has to be taken over uneven ground. Virgil speaks of the 'screech-ing axle' of a heavy cart, and he may well be stating the literal truth.

There is very little evidence for the use of lubricants in antiq-uity, except on racing vehicles, when the friction might be so great as to give rise to a fire risk. Water seems to have been the most commonly used, but it cannot have been very effective.*

*See H. A. Harris, *Lubrication in Antiquity*, Greece and Rome XXI/1, April 1974, 32–36. Prof. Harris was probably right to express scepticism over the supposed 'roller bearing' of the DEJBJERG wagon (Oxford History of Technology, Vol. II, p. 551). The holes might well have been drilled in order to remove the axle-socket, and left there to contain lubricant—perhaps grease.

Wheeled vehicles were of course normally driven on roads or farm-tracks, but there is evidence that in a few places in the Greek world special 'tramways' were built of stone slabs, with deep grooves for the wheels to run in. It is surely not a coincidence that the loads being carried were in each case abnormally heavy—blocks of stone from a quarry or, in one famous case, warships.

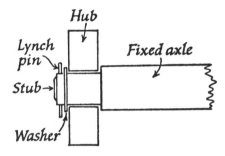

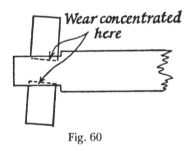

Fig. 60

When remains of such a 'tramway' are excavated, there is naturally some doubt as to whether they were deliberately constructed as 'tramways' or whether they were ordinary roads in which grooves have been worn by the passage of many vehicles over a long period; once a shallow groove had been formed, the wheels of later vehicles would naturally tend to slide into it, and follow it. There are, however, certain features which suggest that the grooves were deliberately cut. They have a remarkably consistent width and depth (20–22cm and 12–15cm respectively) and there seems to have been a 'standard gauge' of 112–144cm. But the most telling piece of evidence is that the grooves are often found to pass across the centres of the stone slabs. It would be a remarkable coincidence if this had happened by accident.

Much the most famous of these 'tramways', the *Diolkos,* or 'pullacross', ran across the Isthmus of Corinth from the Corinthian gulf to the Saronic gulf, and was used by the Corinthian navy to transfer ships back and forth.* It followed the shortest route possible, avoiding steep gradients, and ran not very far from the modern Corinth canal. It is not certain what motive power was used. Because there are no obvious marks of animals' hooves on the roadway, it has been argued that the ships, mounted on wheeled trolleys, must have been manhandled across, probably by their own crews. This argument is plausible, but not conclusive.

In other situations it was necessary to transport abnormally heavy loads overland without using wheeled vehicles at all—where a stone tramway was impracticable or, because it would only be used for a short time during a particular construction job, uneconomical to build. Vitruvius describes ways in which column drums and architraves were transported (X, 11–14). One ingenious method was to use a column shaft or column drum (which would be roughed out round at the quarry) as a roller, by constructing a wooden frame around it, and fixing short iron spigots in each end (by making a socket and pouring in lead) and fixing bearings in the wood frame in which the spigots turned (Fig. 61a). The frame was then harnessed to a team of oxen, and pulled along like a heavy road-roller. This device was invented by Chersiphron, architect of the temple of Artemis (Diana) at Ephesus.

His son Metagenes showed similar resourcefulness. The architrave blocks were of comparable weight but square or rectangular in cross-section, so they could not be rolled along. Accordingly, two wheels were made about 12ft (3.66m) in diameter, probably of the cross-bar type (Fig. 6lb), with heavy bars and broad felloes, so they did not easily sink into the ground. The ends of the architrave blocks were 'enclosed' in the wheels (probably between the cross-bars) and fitted with a spigot at each end which, as with the column drum, turned in a bearing mounted in a wooden frame. The spigots would have to be lined up with the centre

*See B. Ashmole, Architect and Sculptor in Classical Greece (Phaidon Press 1972), pp. 20–21.

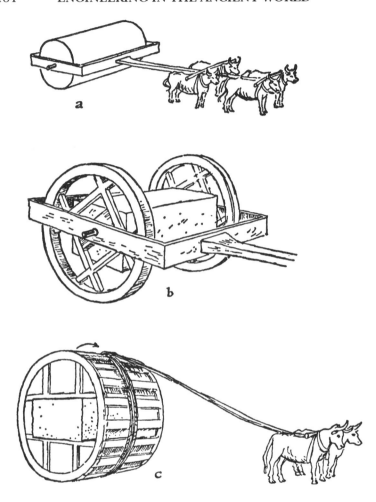

Fig. 61. Transport of column drums and stone blocks.

of gravity of the block to allow the wheels to turn steadily. Vitruvius stresses that the whole operation depended on two conditions, that the quarries were not far away from the building site ($7\frac{1}{2}$ miles, 12km) and the ground in between was flat and level.

Another engineer in Vitruvius' own day, Paconius, tried to go one better, and paid the penalty for his ambition. He had to convey a block of stone 12ft × 8ft × 6ft (to replace a statue base) from the same quarries over the same distance. This volume of stone (576cuft, 16.32m^3) would have weighed something in the region of 45 tons (45.7 tonnes). Paconius used the same method of mounting the block in two wheels 15ft (4.57m) in diameter, but the method of traction was different. He fixed battens of wood two inches (5 cm) thick (which sounds incredibly slender) from wheel to wheel 'in a circle around the stone, at a distance of less than a foot apart'. Vitruvius does not say whether these were on the rims of the wheels or in towards the centre. Paconius then wound a rope around this 'bobbin', and hitched it to the team of oxen. As they pulled, the rope unwound and caused the 'bobbin' to revolve, thus rolling the wheels along the road (Fig. 61c). This is quite a sound method, as it gives a slight mechanical advantage; unfortunately, however, there was a snag. Oxen harnessed to a frame by the older method were able to steer the wheels along the road, but Paconius' structure could not be controlled in that way, and kept swerving to one side or the other. It had then to be pulled back on to the track again. In the end the job took so long that Paconius ran out of fodder for his oxen, and went bankrupt.

This chapter began by contrasting the speed and effectiveness of sea transport with the slowness and clumsiness of land transport in the classical world. Two Roman poets expressed this contrast very aptly in two short poems, the second being a direct and witty parody of the first. Catullus wrote (poem IV) of the small vessel which had brought him home to Italy from Asia Minor:

> My friends, that little yacht you see over there
> Claims to have been the fastest vessel in the world;
> There was no keel afloat, cutting through the water,
> But she could overhaul it, whether called upon to fly
> With oars a-splashing or with linen sails aloft. . .

Among the works attributed to Virgil's early years is a short poem about a transport contractor in the area of his birthplace near Mantua:

> My friends, that man Sabinus you see over there
> Claims to have been the fastest mule-man in the world;
> There was no racing curricle, hurtling along the road
> But he could overtake it, whether called upon to fly
> Post-haste to Town, or down the muddy lane to Brixia . . .
>
> (Catalepton X)

8

The progress of theoretical knowledge

Since the Roman contribution to technology, though consider-able, was almost entirely in the field of practical application, the state of Greek theoretical knowledge may be regarded, for all intents and purposes, as that of the whole Mediterranean world and the Roman Empire, down to the fifth century A.D. or even later. It is therefore unnecessary (and would in any case be far beyond the scope of this chapter) to review the subject as a whole. It will suffice to highlight one characteristic feature of Greek thought, and then to assess the state of theoretical knowledge in three particular areas which are closely related to the main top-ics dealt with elsewhere in this book—hydrostatics, mechanics and chemistry.

It is possible to see, in almost every branch of Greek literature, a particular trait of the Greek mind which had important effects in some branches of scientific thought. It was a liking for stability, rest and permanence, and a corresponding dislike, almost a mis-trust, of change, movement and what they called *genesis* and *phthora,* 'coming-to-be' and 'passing-away'. Why this should be so is something of a mystery, but perhaps their very acute aware-ness of the impermanence of physical things in their world, and of human life itself, caused them to set a high value on the perma-nent and the stable. However that may be, one result was that their understanding of static conditions (e.g. hydrostatics, or mechanical problems not involving movement) was very acute, whereas their ideas on dynamics and ballistics were surprisingly incomplete and inaccurate. They spoke of velocities, relative ve-locities and resistance, but hardly even began to study accelera-tion or deceleration, and they had only a rather vague notion of inertia or kinetic energy. They observed that a stone continued to fly through the air after it had left the hand of the thrower, but throughout antiquity they continued to offer some quite absurd explanations of why it should do so.

When the Greek philosophers first began to enquire into the physical nature of the universe, it was precisely the problem of change and of 'coming-to-be and passing-away' that preoccupied them and, perhaps, disturbed them. More than a century later Plato was still bothered by the same problems, and though his thought was untypical in several important respects, on this point it accorded completely with the Greek tradition. He went so far as to say that physical objects, because they undergo perpetual movement, change and destruction, cannot be 'known' or 'understood' in the true sense of those terms, and one of the aims of his famous theory was to supply, in the 'Forms' or 'Ideas', eternal and unchanging objects of true knowledge, by relation to which material things could be studied and reasonably interpreted, though never truly 'known'. One consequence was what might be termed an 'anti-physical' trend in Plato. Though his *Timaeus* is devoted to a detailed (if rather curious) analysis of the physical world and its creation, he elsewhere exhorts the true philosopher to turn his back on such matters, and 'rise above them' in the pursuit of true wisdom and knowledge. Plato's admirers see this as the honest logical conclusion from his theory of Forms and his epistemology. His enemies see it as the snobbish contempt of an aristocrat and man of independent means for those of his fellows who had to deal with physical objects all their lives, and worked for their living.

This feature of Plato's thought would be less important were it not for the fact that some of his followers in later centuries not only accepted the 'anti-physical' attitude, but carried it even further than Plato had thought fit. Uncritically, and without really understanding his arguments, they exalted the 'pure' and theoretical sciences (such as geometry and astronomy) and looked down on any research that was mechanical, or which had practical applications. Accordingly, if they admired a particular scientist or thinker, they tended as a matter of course to attribute to him all their own prejudices. Plutarch, writing in the first century A.D., is guilty of this in speaking of Archimedes, and the much-quoted passage in which he does so (*Marcellus* 17, 3–5) deserves to be treated with the utmost scepticism.

There were other reasons for the lack of progress in the study of motion. Much the most important of these was the absence of devices which could be used to measure short intervals of time, of the order of a few seconds. They had water-clocks of various types, some of them capable of running for up to 24 hours or more, but when they were used for scientific measurements (e.g. in astronomy) they did not measure the passage of time as such, nor were they calibrated in time units. They measured one time interval as a multiple or fraction of another, such as the time between the first appearance of the sun at dawn and its complete clearance above the horizon, as a fraction of the whole diurnal cycle. Yet another obstacle to research was their method of dividing that cycle into 'hours'. Instead of having 24 hours of equal length, they divided the day, measured from sunrise to sunset, into 12 equal divisions, with the result that their 'hours' were not fixed units of time, and varied from what we would call 'about 40 minutes' in winter to 'about 80 minutes' in summer. This system was adequate and indeed rather useful for the time-measuring needs of everyday domestic life. The working day for the ancients ended effectively at sunset, and, if one had to get through one's daily duties in less time in the winter, it was really just as sensible to have shorter hours as to have fewer hours. But for the astronomer it created serious problems. If he wished to fix a point of time at which an eclipse had been observed, he had to specify the 'hour' and also the exact time of year, and do quite a lot of arithmetic to correlate two points in time fixed by this method. It was probably for this reason more than any other that our system of fixed equinoctial hours eventually replaced the 'classical' system. Until that occurred, it was virtually impossible to subdivide the hour into minutes and seconds, which would likewise have varied in length according to the time of year—a truly daunting idea.

There was, however, one smaller unit of measurement, used in the law-courts of Greece and Rome, which was to some extent fixed and regulated by law and did not vary with the seasons. In order to ensure fairness, the prosecution and defence were each allowed the same amount of time for the presentation of their cases. The measurement was made by the simplest form of waterclock, called a *clepsydra*. It was a jar of fixed capacity with a hole at its base of fixed diameter, through which the water ran away. The time was assessed in terms of 'two clepsydras' or 'five clepsydras', and so on, according to the complications of the case or, in a civil suit, the amount of money at stake. There may have been some attempt to standardize this unit of time, since the surviving remains of one *clepsydra* (in the Agora Museum at Athens) bear two marks which seem to indicate a check on its capacity by an inspector of weights and measures. A reconstructed model of this clock runs for about 6 minutes per filling.

But this use of a time-measuring device was in the interests of fairness, not in the pursuit of science. The same might be said of another application of the same device, recorded in an anecdote concerning a well-known prostitute in fourth-century Athens, who was known professionally as 'The Clepsydra' because, in order to ensure a fair distribution of her favours, she installed a water-clock in her boudoir, and timed her clients' visits by it.

The science of hydrostatics was given its theoretical basis by Archimedes, in the latter half of the third century B.C., as set out in his surviving work, *Peri Ochoumenōn*—'On Floating Bodies'. Of course, a good deal of practical application had been going on for many years on such things as siphons, water-clocks and buoyancy devices, but Archimedes codified the theory, and gave it a sound mathematical basis.

The treatise begins with a basic supposition about the nature of liquids, which deserves quotation in full: 'Let it be assumed (i.e. accepted without proof or demonstration) that the nature of liquid is such that particles on any one level will displace (literally, "shove aside") other particles which are on the same level but under less pressure: and that the pressure is determined by the liquid which is perpendicularly above each particle, provided that the liquid is not in an enclosed container, and under pressure from some other force (i.e. other than gravity).'

He goes on immediately to make clear that by 'on the same level' he means 'at the same distance from the earth's centre', accepting without comment the conclusion reached by fourth-century astronomers that the earth is spherical, and that the force of gravity acts towards its centre. These ideas, so long familiar to us, cannot have been obvious to common sense, or easy to assimilate. Virtually all the rest of the treatise consists of deductions from these basic assumptions, the favourite type of argument being the *reductio ad absurdum*—'if this proposition is untrue, we shall be led to a conclusion which is contrary to our original assumption'.

He first shows that the surface of any mass of liquid in a stable disposition is a sphere, or segment of a sphere, concentric with the earth. The proof is simple. If the surface is anything other than spherical and concentric with the earth, there will be particles on the same level which are under different pressures, and movement will result. So the disposition will not be stable. In contrast to this straightforward logic, there is evidence, recorded by Pliny (*Nat. Hist.* 2, 65, 164), that the question was tackled empirically. The curvature of such a surface is, of course, not discernible except over an expanse of several miles, but Pliny points out that a shining object at the top of a ship's mast remains visible from the shore after the hull has disappeared below the horizon.

Archimedes then (I, 3) passes on to deal with three categories of body, those whose density is (a) equal to, (b) less than, (c) greater than, that of the liquid in which they are immersed. A body less dense than the liquid will float, and the portion of it which is submerged bears the same ratio to the whole volume of the body as its whole weight bears to that of an equal volume of the liquid. If it is forcibly held below the surface it will exert an upward thrust equal to the difference between its own weight and that of the liquid it displaces This thrust was used by Ctesibius to work an automatic water-level control in the water-clock (discussed below), and it fulfils the same function in the ball-cock of a modern water cistern. A body whose density is greater than that of the liquid in which it is immersed will sink to the bottom and, if weighed while submerged, will appear lighter than its true weight by an amount equal to that of the liquid it displaces.

In addition to these theoretical principles, we know from other sources that Archimedes devised a practical method of using them to assess the proportions of gold and silver in a crown made for the king of Syracuse. To do so, it was necessary to measure the exact volume of the crown, and his discovery of a method for doing this must surely be the most famous story in the history of science. On stepping into an over-filled bath-tub, he saw that the water which overflowed, if caught and measured, would give the exact volume of an irregularly shaped body—namely, his own. In his haste to get home from the public baths and try this out on the crown, he made his well-known 'nude dash' through the streets of Syracuse, with shouts of *'Heurēka, Heurēka'* ('I have found it'). This must have mystified the onlookers, who probably were thinking that exactly the opposite had occurred.

Next in the treatise comes a series of problems in stability—the tendency of a floating body to assume a particular position, and to return to that position if tilted away from it and then released. Apart from two propositions on segments of a sphere (I, 8 and 9) they all deal with one particular shape—the paraboloid (the solid traced out by a parabola, which Archimedes calls a 'section of a right cone', revolving on its axis—Fig 62). Its stability is assessed

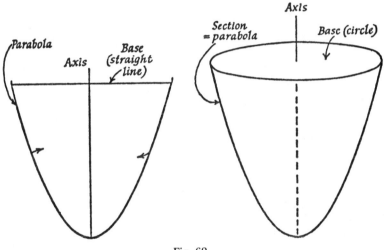

Fig. 62

in relation to two factors, (a) its shape—the proportion of its height to its breadth, which is measured in a very sophisticated way—and (b) its density or, as we would say, its specific gravity. Why Archimedes should have been so preoccupied with this particular shape is not clear; but two explanations have been offered. One is that seen in cross-section it approximates very roughly to the shape of an ancient ship's hull, and that he was thinking, in the very long term, of calculations which might be useful to a naval architect. The more likely alternative is that, having wrestled with various problems connected with the parabola, and having brilliantly solved two of them (measuring the area of a segment and finding its centre of gravity) he felt a kind of affection for this type of curve, and for the solid which it generates.

In contrast to this type of method, we have a few remains of hydrostatic theory in which arguments of a kind which might be called empirical are used. They are preserved in the introduction to Hero's *Pneumatica,* and they include one very striking example of a truly scientific method. Though this is a matter of controversy, there is no real proof that the idea for the experiment was not Hero's own.

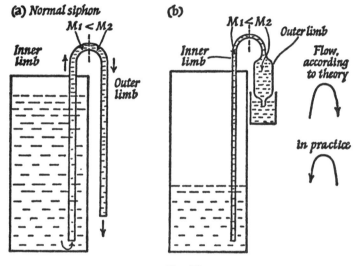

Fig. 63

He has just given an explanation of why a liquid flows through a siphon, which he presumably found in some earlier writer, and which goes like this. If the pipe is of uniform diameter there is a greater volume, and therefore a greater weight of water in the longer (outer) limb (Fig. 63a). This 'overpowers' the lesser weight in the inner limb, and draws it up (on the analogy of a pair of scales). Since air cannot get into the inner limb, and a continuous vacuum cannot be formed, liquid from the vessel must flow in to replace that which is drawn off, and so on. 'But', says Hero, 'we can demonstrate that this explanation is wrong. Construct a siphon with the inner limb long and slender, and the outer limb shorter and much thicker, so as to contain a greater volume of liquid than the inner' (Fig. 63b). According to the theory, the greater weight ought to fall, and raise the lesser weight in the inner limb. Then comes the crucial sentence: 'But this does not actually happen; therefore the explanation I have quoted is not the true one'. All the elements of modern method are here; the formulation of a theory, the use of an experiment (with specially designed apparatus) to test it and, most important of all, the acceptance of the experiment as a conclusive proof that the theory is wrong. (What actually happens is, of course, that the liquid flows 'the wrong way' through the siphon.) There are very few examples of experimental method in Greek science, but it is not true to say that there are none at all.

Both the Archimedean logic and Hero's more empirical approach do, however, share one feature in common—the Greek trait discussed at the beginning of this chapter. On Archimedes' basic assumption, if a liquid is in an unstable disposition (particles on the same level under different pressures), movement will take place and will continue until a stable disposition is reached; but he never, apparently, tried to analyse that movement, or to assess its speed or changes in its speed. He was interested only in the situation of potential movement before it started, and that of zero movement after it had ceased—in other words, the static aspects, not the dynamic. Hero goes a little way in that direction, but not far. He describes the conditions under which a siphon will run, and those under which it will stop running, and he also points out that the rate of flow through it is not constant, but diminishes as the system approaches the 'stop' conditions. But he makes no attempt to work out a mathematical formula for this change in the rate of flow. It had caused his predecessor Ctesibius certain problems in connection with water-clocks. The diminution in the rate of flow also occurs, as Hero points out explicitly, in 'a vessel with a hole in the bottom', which must mean a *clepsydra*. As the water level falls, the rate of flow through the spout slows down, with the result that one cannot measure fractions of the time unit, such as 'half a clepsydra' or 'one-eighth of a clepsydra', by filling the jar half full, or one-eighth full, as the case may be.

Ctesibius' solution of the problem did not involve tackling the dynamics of the system. He merely stabilized the rate of flow and made it constant—in fact, he took the dynamics out of it. Instead of measuring the fall of level in an emptying jar, he kept it constant with an automatic valve worked by a float, similar to the needle valve in the float chamber of a modern carburettor (Fig. 64a).

This constant outflow was used to fill a straight-sided vessel, and the rise of level in that vessel could be measured, and subdivided into fractions as required. Hero's solution, on the same basis, was to construct a siphon with a float on its inner limb, and a mounting which enabled the siphon as a whole to rise and fall with fluctuations in the water level (Fig. 64b). As he points out, the rate of flow depends on the 'head' of water—the difference of level between the surface of the liquid being drawn off and the outlet of the siphon which, by this arrangement, is kept the same at all times.

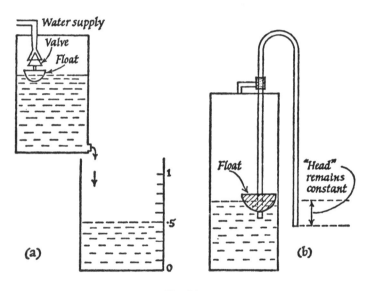

Fig. 64

Our information on Greek theoretical knowledge of mechanics comes almost entirely from two works, *Mechanical Problems,* attributed to Aristotle but probably of later date, written by a member of his school, and Hero's *Mechanica,* discussed in Chapter 9.

Mechanical Problems cover some of the same ground as Hero's *Mechanica*—pulley systems, the wedge and the lever. At the start, the author plays about with certain rather paradoxical thoughts about circles. A typical example is that if a radius revolves, the points along it move together and in a line with each other, and yet those further out from the centre cover a greater distance than those nearer the centre. This leads onto a number of principles of levers, which are conceived in the 'classic' form of mover, force, lever, fulcrum and load (Fig. 65a). The longer the lever (between mover and fulcrum), the greater the force exerted on the load, and the weight of the load and the force needed to lift it are inversely proportional to their distances from the fulcrum (Fig. 65b), and so on.

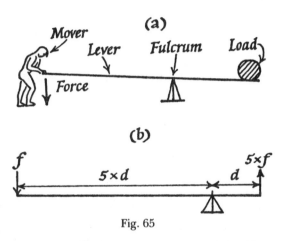

Fig. 65

A number of attempts are made to use this as a model to interpret various mechanical systems, several of which involve ships (probe. 4–7). The oar is seen as a lever, the rowlock as a fulcrum and, quite correctly, the sea as the 'load' which is moved. Moreover, this movement of the load causes a 'counter-thrust' on the fulcrum which propels the ship forwards. When we come to the rudder (prob. 5) the model does not work so well. The point at which the rudder is fixed to the ship is the fulcrum, the whole rudder is the lever, the sea is the load and (here he goes astray) the helmsman is the moving force. Once again, because the 'load' is thrust in one direction by the 'lever', the fulcrum (and with it the stern of the boat) is thrust in the opposite direction. The writer is groping towards the Newtonian principle that 'action and reaction are equal and opposite' without actually arriving at a formulation.

Another of the ship problems (no. 4) is a telling example of the static conditions being understood, but the dynamics viewed wrongly. It is stated that the rowers amidships contribute more to the propulsion of the ship than those near the bows or stern, because they sit further inboard, and have a longer 'lever' (i.e. the oar between the hands and the rowlock). They therefore exert a greater force on the 'load', and though this is quite true, the fact that their oar-blades move more slowly is ignored. The lever model breaks down completely in prob. 6, where it is used to account for the fact that if the mast, yard and sails are higher, the ship will sail faster.

Problems 8, 9, 10 and 11 deal with wheels and rollers, and show some understanding of the problems of friction which arise in designing carts and wagons, but the writer's preoccupation with circles and leverage leads him sometimes to omit the obvious— e.g. that a large wheel turns more easily than a small one because it does not go so far down into the potholes.

The nearest approach to a concept of kinetic energy comes in prob. 19. Why is it that, if one places an axe on a log of wood, and a heavy weight on top of the axe, it will not split the wood, yet, if the axe is swung down onto the log it will split it, even though the axe is lighter than the weight which was placed on it? 'Is it,' he asks rather tentatively, 'because all work is done by means of movement, and one weight is capable of imparting more movement to another if it is moving than if it is stationary?' Once again, an important concept is hinted at but not explored in detail.

The last four problems deal with motion, and show the author's characteristic Greek shortcomings. Indeed, in prob. 32 (if the text is correctly restored) he seems to suggest that there is no point in exploring such problems, because they are insoluble. From the variety of alternative explanations he gives for the deceleration of a missile, it is clear that there was no single agreed theory. In the final problem, however (no. 35) his preoccupation with the proper ties of circles leads to a fundamentally correct explanation of a puzzling phenomenon.

Several of the early philosophers had suggested that the universe was created from an original state of chaos in which all kinds of matter were mixed up at random, by a circular motion, or 'eddy'. This motion caused the lightest particles (those of a transparent, fiery substance called *aether*) to fly outwards to the periphery and form the sky. The next heavier kind, those of air, took up a lower position, those of water came next, and the heaviest of all, those of earth, came together at the centre. At first sight, a modern scientist might take this as curious reversal of the truth, since the heavier particles, having greater weight, and thus subject to a greater centrifugal force, would be more likely to fly out to the periphery than the lighter ones. But it must be remembered that the Greeks had no means of investigating the behaviour of particles moving in a vacuum. They thought of the 'eddy' as a movement of all matter, mostly air, which carried the heavier particles around with it. The illustrative model they had in mind was, in fact, a very homely one—a bowl of soup with lumps in it of various sizes and weights. If set in motion it would, as it 'eddied' around the bowl, exhibit exactly the behaviour which the philosophers assigned to the universe—the biggest and heaviest lumps (of turnip, perhaps) would gather at the centre first, and the lighter ones towards the outside. This is how (at the second attempt) our author explains what is happening. 'Or is it because those particles which the eddying water cannot control, because they are too bulky and heavy, must inevitably be left behind by it, and travel more slowly?' He goes on to explain that if they move inwards to a smaller orbit they will in fact be travelling more slowly, although in a sense all the 'circles' are revolving at the same speed (as we would say, their angular velocity is the same). The result is that they keep on moving in to a smaller orbit until they reach the centre.

The Greek achievement in chemistry presents us with an odd contrast. On the one hand, Greek doctors knew and used a very wide range of chemicals, both mineral substances and others extracted from animals and plants by various processes. But on the other, there was virtually no general theory about chemicals, or the laws which govern their combination or separation. The explanation for this must lie in the fact that the two types of knowledge are acquired in very different ways. The ancient doctor could try out any one of the collection of chemicals in his cupboard (from which, of course, the known harmful poisons had been excluded) and see if it helped his patient's condition. If it did, he could then prescribe it for other patients with similar complaints, and if it did not, he could try another. (This procedure is rare in modern medical practice, though not altogether unknown.) But from start to finish, he did not really need to know the chemical content of his medicines, or to understand the real reasons why they were beneficial in some cases. All he needed to know was the source of the material, the method of preparing it and (closely related to that) the dosage. One striking example illustrates how far this empirical method could take them, and how the lack of techniques of chemical analysis and testing prevented them from going any further. Dioscorides (first century A.D.) the author of the best known ancient work on *Materia Medica,* mentions a drug prepared from *mandragora officinalis* which, he says, can be used as an anaesthetic. We have no other evidence for its being so used, probably because there was no means of testing the strength of a dose, which had to be accurate within very narrow limits.

In fact, chemical testing and analysis was, for the Greeks, almost entirely a matter of using the senses. Of the range of terms they used for what we would now call chemical properties, there is hardly one which does not relate to taste, smell or touch. When they speak of unripe fruit being 'sharp' (*oxy*) it is tempting to translate it as 'acid', but if we do, we must forget all about litmus paper or any other chemical reaction. All the Greek word means is 'something which tastes like vinegar'. Incidentally, there was no word for alkali, simply because they did not regard alkaline substances as being linked by a common taste factor.

Their understanding of the chemical effects of heat was also imperfect. Because a number of familiar chemical processes are speeded up by an increase in temperature, they tended to think that the heat was itself the cause of the process, and not merely one of the controlling factors.

However, when all this has been said in disparagement of their chemical theory, it is as well to recall that the Greeks and Romans managed their practical chemistry with fair success. They could cure some diseases; they could brew palatable and potent wines; they could devise an adequate and healthy diet from terribly limited resources of food production; and they could put a black glaze on their pottery which still looks shiny and new today and—who knows—may still remain so when our own civilization has passed into history.

The principal Greek and Roman writers on technological subjects

SOME of the most difficult problems in the study of Greek and Roman technology arise from the fact that our information comes from a wide variety of sources. Odd references to such subjects as metallurgy, transport or building technology are scattered about in the works of many authors, and the archaeological evidence comes from a large number of sites, and ranges over a long period of time.

A complete review even of literary sources, therefore, would be a very lengthy and tedious undertaking, but there are three ancient writers, Hero of Alexandria, Vitruvius and Frontinus, whose works (or some of them at least) are wholly concerned with technological subjects, and these writers will be considered in some detail. A fourth, Pliny the Elder, compiled a great encyclopaedia in which technological subjects are referred to (along with almost every other subject under the sun), so he deserves at least a brief mention.

HERO OF ALEXANDRIA

Hero (the Greek spelling of his name was Heron, but the Latinized form is more commonly used) is one of the most important sources of our knowledge about ancient technology. His works have survived in some quantity (though they are not complete), and he was a very versatile man, with the result that he supplies information on pure mathematics, physics, mechanics, conjuror's apparatus, surveying instruments and many other items plus, most interesting of all, some occasional insights into practical engineering at the 'nuts-and-bolts' level. A number of his devices have already been discussed in the appropriate chapters.

Of the man himself we know virtually nothing apart from what can be inferred from his writings. His name suggests that he lived and worked at Alexandria; several later writers also referred to him as 'Hero the machine-man' *(mechanikos)*. Of his social status we know nothing, except that he was an educated man, well-read in the works of the mathematicians and engineers, particularly those of Ctesibius, the most famous in the ancient world. Hero's position, however, was probably very different from that of Ctesibius, who lived in the third century B.C. when Egypt was ruled by the Ptolemies, hereditary kings descended from one of Alexander the Great's Macedonian generals. They were extremely wealthy and are known to have patronized both literature and science. A number of surviving technical works are addressed (either as a courtesy to a present patron or in the hope of attracting new patronage) to them and to other contemporary Hellenistic rulers, who could offer similar, though perhaps less lavish, benefits. For instance, Biton's work on *Engines of War and Catapults* was written for King Attalus of Pergamon. But if Hero lived in the first century A.D., as will be argued in the next paragraph, the Alexandria of his day was under Roman rule. Cleopatra, the last of the Ptolemies, came to her much-dramatized end in 30 B.C., and Egypt thereafter was kept under very strict control by the Romans, being one of their most valuable sources of imported wheat. On the other hand, there is nothing in Hero's writings to suggest that he worked for the Roman government or for a Roman patron, as many of his Greek contemporaries did. He sometimes uses Roman names for weights and measures transliterated into Greek, but that was common in legal documents and business agreements made in Egypt under Roman rule. He also mentions Latin equivalents for Greek technical terms, which seems to suggest that he was acquainted with some Latin technical literature (which is surprising), and that he hoped to find some readership among those Roman engineers who could read Greek.

The question of his date has been much disputed, suggestions ranging from the second century B.C. to the second A.D. or even later. The fact that he recognized the existence of the Romans at all would seem to rule out a date earlier than mid first century B.C., and makes a later date much more probable. The question was very open until in 1938 the machinery of modern science was brought in. In the *Dioptra* Hero describes a method of calculating the Great Circle distance from Rome to Alexandria by observing the same eclipse of the moon in both places, and measuring the time interval between (Chapter 35). Though he seems to be talking in hypothetical terms ('let us suppose that an eclipse has been observed . . .') he chooses a date 10 days before the spring equinox. It is surely not a coincidence that between 200 B.C. and A.D. 300 the only eclipse which met the required conditions (visible both from Rome and from Alexandria) occurred, according to modern astronomers, on March 13 A.D. 62. Though this does not prove (as has been asserted) that Hero was alive then and recorded the event in his *Dioptra,* it does show that he knew of an attempt to calculate the distance which was made on that occasion, and that his *Dioptra* must date from later than A.D. 62. Since the geographer Claudius Ptolemaeus, writing about the middle of the second century A.D., uses slightly more sophisticated methods, we may safely assume that Hero was active (a not inappropriate term) between about A.D. 50 and A.D. 120. This is also consistent with the probable date of his *Cheiroballistra,* discussed in Chapter 5. It was the age of the Flavian emperors at Rome, Vespasian, Titus and Domitian, and their successors Nerva and Trajan.

Hero's surviving works are here listed in the order of the standard edition in the Teubner Library.

(1) *Pneumatica.* There is no really satisfactory English translation for this title. '*Pneumatics*' will hardly do. Some of the devices described do use compressed air, but others use siphons and steam pressure. The German 'Druckwerke' translates it exactly, so perhaps 'pressure-mechanisms', if clumsy, is the nearest one can get.

The work, in two books, begins with a theoretical discussion of the properties of air, the impossibility of a continuous vacuum, and the behaviour of liquids acted upon by gravity. Hero is clearly dependent on earlier writers for this section, but his originality consists in drawing together two schools of thought with very different styles and interests. The first part is derived from Strato of Lampsacus, who was the third principal of Aristotle's school, the Lyceum, from about 288–268 B.C. In keeping with the Aristotelian tradition, he tends to be descriptive without quantification and hence almost completely without mathematics, and he also uses empirical arguments and demonstrations to explain or corroborate his theories. But in the second part of the introduction, and in Chapters 1 and 2 where he is discussing siphons, Hero draws on Archimedes, who started, not from empirically demonstrable ideas, but from certain assumptions (for which he offers no proof) about the nature of liquids. His deductions are all strictly quanti-fied, and in some cases highly mathematical.

After the introductory matter, Hero describes some 75 structures or devices, the great majority of which are ornamental and enter-taining rather than useful. Many of them are for dispensing wine, or mixing wine and water, including a trick jug (I, 9) with two concealed air inlets, from which one can pour at will wine, or a mixture of wine and water, or 'if we want' (he says) 'to play a joke on somebody', water. Other devices include an organ with constant air pressure maintained in a hydraulic reservoir, a pump used for fire-fighting, a fountain worked by compressed air and a jet-propulsion steam engine. They also include some devices con-nected with temples and religious observance, and inevitably our moral attitude towards Hero is closely bound up with our inter-pretation of these devices and of Hero's motives in designing them, which has been a matter of much contention among scholars.

The late Benjamin Farrington, in his *Greek Science**speaks scath-ingly of those Alexandrian scientists (unnamed) who worked for the Ptolemies, and whose science 'became the handmaid of reli-gion and was applied to the production of miracles in the Serapeums and other temples of Egypt' (p. 199). He seems to imply, without ever saying so, that Hero was employed in the same disreputable line of business ('The scientific production of mira-cles covers the whole period of the rise and fall of Alexandrian science', p. 200)—but in the service of what government, or in what cult context, he does not say. What were these 'temple miracles'?

Of the 75 devices in the *Pneumatica*, 11 have been, or might be, listed in that category. Two of them (I, 14 and 23) have siphon systems by which, when water is poured into one jar, wine comes out of another. This trick could scarcely have aroused much 'be-wilderment and awe'; for it to have any hope of qualifying as a 'miracle', the wine would surely have to flow from the same jar. Four more can be effectively ruled out, because they were mini-ature scale models—play-things of the idle rich, which could not have had any widespread influence on the general public who are most unlikely to have seen them. These are I, 12, II, 3 and 21, plus I, 38 and 39, two methods of making miniature temple doors

*Pelican Books, 1953.

open when a fire is lit on a toy altar. This leaves us with five actual 'temple miracles'.

I, 17 is a device for making a toy trumpet play a single prolonged note when a temple door (full-scale this time) is opened. How profound a reverence this would inspire in a worshipper is doubtful, especially if he spotted the tell-tale piece of string running from the top of the door over a pulley. II, 9 describes a thyrsus (a wand with imitation leaves at one end, emblem of the Greek god Dionysus) which whistles when thrust into water. Again, we can hardly assess the emotional effect with confidence, especially on non-Greeks. I, 21 is a coin-in-the-slot machine to be placed at the entrance to a temple which, in return for a five-drachma piece, dispersed a small amount of water for ritual washing of the face and hands. No doubt worshippers would entirely accept the need to purify themselves before entering a holy place, but their thoughts on being confronted by a sort of one-armed bandit, which stung them several days' wages for the privilege, might not have been altogether devout. Finally, Hero explains in two places (for the benefit of Romans unacquainted with Egyptian customs) that some Egyptian temples had bronze wheels mounted beside the entrance, which the worshippers turned round as they passed in, 'in the belief that the bronze purifies them'. He suggests two improvements on this device. One (I, 32) is a valve which causes water to spout from the hub of the wheel when it is turned, thus streamlining the two purification rituals (by bronze and by water) into one. The other, his ultimate essay in fiendish ingenuity (II, 32), was to mount the bronze wheel on the side of a box, on top of which was a little stuffed bird which, when the wheel was turned, spun round on a vertical axle and warbled. So much for the 'temple miracles', and for science in the service of religion and oppression. Hero's other surviving works may be treated more briefly.

(2) *Automatopoiētikē* ('the making of automata') an account of the construction of two miniature mechanical puppet theatres. Being designed long before the clockwork age, they are powered by weights in the form of pistons which drop into cylinders. Before the start of the performance, the cylinder is filled with millet or mustard seeds, the piston resting on top of them. As the seeds run out through a hole in the bottom of the cylinder the piston sinks slowly down, pulling on a cord which turns the main shaft and

activates what Hero calls the 'plot of the play' *(mythos)*, and we would call the programme. In the first device, the whole apparatus moves forward on wheels and stops in the right position; figures then revolve or move about, doors open and close, fires burn up on tiny altars, and so on. At the end it moves back again out of sight. The second device does not move as a whole but has more figurines, performing a greater variety of movements, some of them (rather oddly, in view of the likely audience) representing artisans at work. As it requires less power, dry sand is used instead of millet seeds in the cylinder, since it runs out more slowly, and enables the programme to last longer.

Though these devices are of course strictly for entertainment, they raise a number of interesting problems of design. They both work very close to the limits of the power available, and must be well designed and skilfully made if they are to run at all. Hero specifies with great care the materials to be used. For example, the timing of the later items in the programme is managed by leaving various lengths of slack in the cords which operate them, and these, therefore, must not shrink or stretch, otherwise the sequence will go wrong. For this reason, he says, gut strings must not be used, because they are affected by air temperature and humidity, a fact well known to musicians, both ancient and modern.

It has often been remarked that Hero, though he devised some ingenious working models, never thought to apply the 'automation principles' used in these models to full scale industrial use. In reply to this two points must be made. The power source used in the models was very clumsy and feeble, and could not have been made effective on a larger scale. Imagine, for instance, a weight of 550lb (250kg) hoisted to a height of 13ft (4m) with a block-and-tackle. With its rate of fall suitably controlled, it could do one man's work (that is, develop 0.1 h.p.) for just over two minutes, and would then have to be hoisted up again. It is, of course, not really a power source at all, but a device for storing energy, and it would be much simpler and more efficient to use man-power to work the machine directly, instead of hoisting weights. Secondly, 'automation' is a relative term. In the second device the arm of a figurine representing a blacksmith holding a hammer is made to rise and fall by a sort of cam and lever arrangement similar to that by which the windmill in *Pneumatica* I, 43 pumps the organ

(p. 26). In order to reproduce this on a larger scale, a different power source would be needed. Water-wheels may have been used some three centuries later to work stone-saws in this way, but Hero was apparently not acquainted with them. And even if the hydraulic hammer, worked by a water-wheel, had been developed in antiquity as it was in the eighteenth century, it would be a little misleading to call that 'automation'. The term nowadays usually implies the complete replacement of human skill as well as human physical effort, and to use a hydraulic hammer, which thumps regularly with the same force on the same spot all the time, is by no means an unskilled job.

(3) *Mechanica.* The original Greek text of this work, apart from a few short excerpts, does not survive, but we have a version in Arabic, translated from the Greek in the mid-ninth century A.D. by a scholar whose name is usually Anglicized as Costa ben Luka. In addition to the text, this version contains a number of figures, possibly drawn by the translator to illustrate his work, or derived from his Greek text, in which case they might even be Hero's own. The treatment of pespective is odd and sometimes confusing, and later copyists of the manuscripts may have introduced some errors, being perhaps quite ignorant of mechanics, but the drawings are of great interest to the historian of mechanical draughtsmanship, being among the earliest attempts to represent such things as gears, levers and pulleys in two dimensions.*

The treatise is divided into three books. The first (after a description of a geared winch, which is probably misplaced) deals with theoretical principles—gear-ratios and their implications, the parallelogram of forces, the scaling-up or down of various plane figures, using a pantograph with two rack-and-pinion drives in a fixed ratio to each other, and another device for scaling up or down a 3-dimensional figure. Then pulleys are dealt with, and the gearing-down effect of block-and-tackle arrangements, and in the last 11 chapters, closely based on works of Archimedes now lost, there is a study of the location of centres of gravity and the distribution of loads.

---

*A substantial section of A. G. Drachmann's book *The Mechanical technology of Greek and Roman antiquity* (Copenhagen and Wisconsin, 1963) is devoted to this work, with translations of the crucial passages and commentary (19–140).

Book II deals with the basic mechanical devices—the windlass, the lever, the pulley, the wedge and the screw, the last being used either with a slider which engages the thread or with a cog-wheel (worm-and-pinion). Each of these is investigated in turn and then, typically, in various permutations and combinations. The mathematics of worm gears—the pitch and spacing of the threads, the best profile for the groove and the cog-teeth, and so on—are treated in some detail.

Book III deals with practical applications of these basic devices, mostly in cranes and hoists, and presses. This section takes us right into the practical mechanics of the ancient world, and is one of the most valuable for the historian of technology. The descriptions include sledges for transporting heavy stones, cranes of various types—the single-pole jib, the shear-legs and the derrick, presses of various kinds, a very ingenious tool for cutting a female thread (Hero actually uses that term, *thelu* in Greek) in a block of wood, and some devices for attaching cranes to blocks of stone, including the lewis bolt.†

(4) *Catoptrica,* the theory of mirrors. This treatise survives only in a medieval Latin translation, in which it is ascribed to Claudius Ptolemaeus, but was almost certainly by Hero. After a brief review of earlier theories, it discusses the phenomenon of refraction, using the concept of a minimum light path to explain certain effects. Then various types of mirror are discussed, flat, concave and convex, and some special arrangements are described for useful or trick effects.

(5) *Metrica,* in 3 books, is a treatise in pure geometry on mensuration. Book I deals with plane figures, triangle, trapezium, polygon, segments of circles, ellipses, etc. Book II goes on to rectangular solids, spheres and segments of spheres, cones (right and scalene), prisms, polyhedra and others. Book III deals with methods of dividing areas (both plane and spherical) in given ratios.

(6) *Dioptra.* This work deals with the theory and practice of surveying, the opening section being taken up with a detailed description of the surveying instrument which gives its name to the whole work. This is a fairly sophisticated plane table, with sights

†A number of short extracts from the original Greek text of the *Mechanics* have been preserved by Pappus of Alexandria (third/fourth century A.D.), and one passage, on the geared winch, is contained in the Greek manuscripts of the *Dioptra.*

and water-levels, adjusted to the correct alignment by worm-gears. It is used in conjunction with vertical poles, which have marker discs which slide up and down them, to determine the relative heights of various points. From Chapter 6 onwards a series of procedures to meet various surveying problems is outlined; how to take a set of sightings along a water supply route, add up all the rises and falls from one to the next, and calculate the total rise or fall. If there is an adequate fall, a gravity-flow system will be feasible, but if the delivery point is higher than the source, pumps will have to be used, and it is here that he casually mentions several types (multi-bucket, chain, screw, drum) clearly assuming that their design is known to his readers. Other procedures determine the distance between two points not visible from each other, measure range (by a sort of crude triangulation) or locate the starting-points for a water-tunnel on opposite sides of a hill. An apparent resemblance between the shape of the hill Hero seems to be talking about and that of the mountain in Samos through which a water-tunnel was cut in the sixth century B.C. has led some scholars to suggest that he had that particular work in mind; but many hills are very like many others, and there seems little point in choosing as an illustration an engineering problem which had been solved many centuries before. The positions for vertical air-shafts, required at regular intervals along such tunnels, can also be found by the *dioptra*. The next problems concern land measurement and area divisions, and then, in Chapter 34 a *hodometer* (road-measurer) for long distances overland is described. It is activated by a chariot wheel, and has a series of worm gears which turn pointers from which distances up to a hundred miles or more can be read off. For still greater distances, or distances over sea or inaccessible land, an astronomical method of measurement is given, and it is here that the eclipse of the moon, which has been used to determine Hero's date, is mentioned. The last two chapters describe a geared winch, and are clearly remnants of the Greek text of the *Mechanica*, wrongly placed here in our manuscripts.

(7) *Definitions.* This is the opening section of a collection of mathematical extracts made by a Byzantine scholar in the eleventh century A.D., and Hero's authorship is generally agreed but not certain. As the title suggests, it is a long catalogue of terms, most of them geometrical, but some concerned with weights and measures, from which some useful information can be derived.

(8) *Geometrica.* This is in the form of an introduction to geometry, covering some of the same ground as the *Definitions,* but in a rather more fragmented and disorganized state.

(9) *Stereometrica.* This work on solid geometry is more characteristic of Hero's style in the *Dioptra.* He begins with pure theory, the mensuration of spheres, cones, pyramids, etc., but the later problems have a less academic and more practical ring about them. He shows how to calculate the seating capacity of a theatre from the length of the highest and lowest rows in the auditorium. Though the allowance of one foot width per spectator is not exactly generous, the arithmetic is correct. He also calculates the number of jars *(amphorae)* which could be stacked in the hold of a ship of given draught, beam and length, and the number of tiles needed to roof a building of given size and shape. Some of these calculations are very rough-and-ready approximations, to serve an immediate practical need.*

Two other extant works of Hero were, for various reasons, not included in the Teubner edition—The *Belopoeica* (on catapult construction) and the *Cheiroballistra* (some specifications for a 'hand catapult'.) These were edited with a commentary by E. W. Marsden (O.U.P. 1971) and are fully discussed in Chapter 5. A few other works now lost, are mentioned by Hero himself and by later writers, probably the most important being a work on waterclocks and time measurement *(peri hydrion horoskopeion),* the loss of which is particularly to be regretted.

VITRUVIUS

The Latin treatise *De Architectura,* in ten books, is the only work of its kind to survive from the Roman world. It had an immense influence on the thinking of architects and scholars throughout the later middle ages and Renaissance, not least on Palladio, and thus (indirectly) on eighteenth-century English architecture.

But of its author we know remarkably little. The name Vitruvius is a family name (i.e. in Roman terms, the middle one of three). The work is addressed to the Emperor Augustus, and refers to the re-building of Rome after the civil war which ended in the battle of Actium in 31 B.C., and so it must date from after that time. But

---

*_Mensuration_ is another scrappy collection, largely covering the same ground as (7), (8) and (9).

he does not use the title 'Augustus' in addressing the Emperor, and since that title was given to him in 27 B.C., this suggests that Vitruvius' work dates from the four years between 31 and 27 B.C.

Julius Caesar, Augustus' adoptive father, had first given Vitruvius some sort of recognition, but we do not know the details. After Caesar's death (probably during the 10 years or so of civil war which followed before Augustus eventually achieved mastery) Vitruvius was appointed, along with three colleagues, to be in charge of the construction and repair of catapults. In other words, he was in a technical section of the army called the *fabri* ('engineers'), who were responsible for weapon maintenance, bridge building, transport vehicles and other such matters. The senior officer in charge of all these operations was called *praefectus fabrum* but Vitruvius does not claim to have held that post. He also gives another interesting detail, that the influence of Augustus' sister Octavia (who was married for a time to Mark Antony) had helped him up the ladder of promotion.

Otherwise he tells us little. In the preface to Book II (para 4) he says that he is not tall, and age has made his face ugly, but this is in contrast to the youth and elegance of Dinocrates, a Greek architect sponsored by Alexander the Great, and Vitruvius is really suggesting that he himself has to stand on his merits, without any kind of unfair advantage. He may possibly have held some official position under Augustus as 'overseer of works', which carried a salary or pension (he says he 'has now no fear of poverty for the rest of his life'), and according to Frontinus, he was responsible for standardizing pipe and nozzle sizes in the public water supply.*

The title *De Architectura* should be interpreted in a rather broader sense than the word 'architecture' normally bears. The root meaning of the Greek word *architectōn* is not so much 'master craftsman' as 'craft organizer', the man responsible for co-ordinating and directing the work of a number of craftsmen with various different specialized skills, and this is how Vitruvius sees his profession.

He begins the first book with an extended account of the education which he considers appropriate for the would-be architect,

---

*The attempt (by P. Thielscher, in Pauly-Wissowa IXA, 427) to identify him with Mamurra, Caesar's engineer officer, is not to be regarded as successful.

and it certainly looks a formidable programme. He must be literate and able to express himself clearly; he must be a skilled draughtsman, able to draw plans, elevations and perspective sketches; he must be a competent mathematician, particularly adept at geometrical constructions and arithmetic. He must also (an interesting sideline) be a well-read and well-educated man with an encyclopaedic knowledge of mythology and legend, so as to be able to plan such features as pedimental sculptures or friezes without making factual 'howlers'. He must be a keen student of several branches of philosophy, particularly natural philosophy and moral philosophy (this is supposed to arm him against avarice and corruption!); he must also understand the rudiments of acoustics and musical theory, and must have a general knowledge of medicine, particularly as it relates to public health. He must also have a good grounding in the law; he must know the legal precedents for various measures concerning drainage rights, lighting, etc., and must be able to draw up a contract which is clear and unambiguous, and will not give rise to litigation later. Finally, since towns and camps had to be orientated without a magnetic compass, he must have enough basic knowledge of astronomy to work out the directions from the sun and stars, and he is also expected to be able to mount and calibrate sundials at various different latitudes. All this is in addition to his knowledge of the central skills of architectural planning, building design, strength of materials and so on. The work as a whole reflects this very wide range of diverse interests, and it is, as the author himself says, 'a complete encyclopaedia for the architect'.

The remainder of Book I is taken up with some definitions of basic concepts such as 'arrangement' *(dispositio)* and 'proportion' *(symmetria),* and with the approach to the first and basic problem of all architecture, choice of sites. This question is discussed in relation to several factors, the main ones being the direction of prevailing winds, and the availability (near at hand) of building materials.

Book II begins with a brief account (based on common-sense and guesswork, but proved to be quite accurate by modern archaeology) of the earliest stages in building—primitive wattle-and-mud huts. Then, after a short excursus on the nature of matter (one of the great pre-occupations of natural philosophy), Vitruvius lists and evaluates the main building materials—brick (sun-dried and kiln-fired), sand, lime *(calx)*, and *Pozzolana*, a kind of volcanic dust found near the Bay of Naples, which was used as a light, strong and waterproof cement. In Chapter 7 he goes on to the principal types of stone—marble, tufa, sandstone, soapstone, etc. Then (Chapter 8) comes an account of various techniques of wall-building, and in the last two chapters a critical catalogue of various types of timber.

Books III and IV deal with the siting, design and decoration of temples, with particular emphasis on the 'orders', Ionic, Doric and Corinthian. Book V describes the other public buildings in a town—forum, basilica, baths, etc. In the course of his description of the plan for a theatre, he refers to a system of resonating jars for the enrichment of the acoustics.*

Book VI begins with an excursus on climate and its relationship to building design, and then comes an extended account of house design and planning, for town houses, country villas and farm buildings. Book VII is mainly concerned with decoration, internal and external, the preparation of stucco, colouring materials and so on.

The last three books deal with particular problems which to us seem peripheral to the actual business of building, but are none the less important. Book VIII deals with water supplies and engineering, and Book IX with astronomy and optics, and the whole technology of time-measuring devices, sundials and water-clocks. The last book describes a number of mechanical devices—cranes, water pumps, water wheels, catapults and other siege engines.

FRONTINUS

Unlike Hero and Vitruvius, who were concerned with a diverse range of subjects, Frontinus concentrated on a specialized area of technology—water supplies and engineering—but we know rather more about him as a person, and about his career outside the field of technology.

His full name was Sextus Julius Frontinus, so he must have belonged (though probably in an obscure and minor branch) to

*See J. G. Landels—'Assisted resonance in Greek theatres', in *Greece and Rome* XIV/1 (1967) 80–94.

the great aristocratic family of the Julii.* He was born about A.D. 35., and held his first major political office in A.D. 70. He reached the highest rank available to him under the emperor when he became consul for the first time in A.D. 73. He was made governor of the province of Britain for about 3–4 years and may have been responsible for the establishment of a legionary station at Isca (Caerleon). We know little of his activities for the next 20 years, except that he composed a work, which still survives, called *Strategemata*—which is a collection of anecdotes from Greek and Roman history illustrating the value of various tactical measures, a sort of Field Officers' Manual. Surprisingly, it makes no reference to his own military experiences, and is quite different in tone and style from the later treatise on aqueducts.

In A.D. 97 he was given responsibility for the water supply of Rome *(cura aquarum)* by the emperor Nerva. This, of course, is the period of his life with which we are mainly concerned. We cannot say for certain how long he held the office, but he died in A.D. 103 or 104, so it was probably for most of the remaining years of his life. During that period he was honoured with the Consulship twice more, in February 98 and January 100, each time as a 'colleague' (so the polite fiction maintained) of the emperor Trajan.

Frontinus, in fact, presents us with two contrasting images, and something of a problem for the social historian. On the one side we have the patrician, with at least some blue blood in his veins, owning villas near the sea at Formiae and Terracina, and following the conventional career of a Roman aristocrat, via political office and military command. Then, after having attained the highest rank, when he was in his early sixties, he took on a totally different, and apparently much less exalted commission. It is true, as he himself says, that the health and well-being of the whole urban community depended on the efficient management of the water-supply, and he adds that the office had regularly been held by 'some of the most outstanding men in the state' *(per principes civitatis viros)*.

From Frontinus' own work *De Aquis,* particularly the Preface, comes the contrasting image of a man who did not, as a Roman aristocrat was conventionally supposed to do, consider the techni-

*There are more than 500 members of this family listed in the Pauly-Wissowa *Encyclopadie,* of which Frontinus is no. 243 (Vol. X p. 591). In most other works of reference he is listed under F.

cal details of water engineering beneath his dignity. His first action on taking over was to make a detailed personal inspection of the entire aqueduct system (one can imagine the dismay this must have caused among the minor officials) and to compile a short treatise on the essential technical details, primarily for his own use, but also for the benefit of his successors. The reasons he gives for doing so, and for starting the work immediately on taking office, show him as a combination of the conscientious public servant and the shrewd officer with experience of commanding men. 'I have always made it my principle', he says (preface 1–2), 'considering it to be something of prime importance, to have a complete understanding of what I have taken on *(nosse quod suscepi)*. For I do not think there is any other surer foundation for any kind of undertaking, or any other way of knowing what to do or what to avoid; nor is there anything more degrading for a man of self-respect than to have to rely on the advice of his subordinates in carrying out the commission entrusted to him.' Of course, he says, subordinates and advisors are a necessary part of the organization, but they should be its servants, not the masters to whom the senior official in his ignorance has to keep running for advice. This calls vividly to mind a phenomenon which could occasionally be observed in technical corps of the Army in the last war—an officer whose technical knowledge was less that it should have been, and who was for that reason held in very scant respect by the NCOs and men under his command.

What kind of a post was the *cura aquarum,* and how well fitted was Frontinus to take it on? On the technical side, he had no special training. Such knowledge as he shows must have been derived from his own reading, mainly from Greek authors who dealt with the elementary principles. The limitations of that knowledge he shared with virtually all the scientists of antiquity, and it would be unreasonable to expect a Roman administrator to be able to solve problems in which Archimedes himself had apparently no interest.

Frontinus' military experience would be of great help in preparing him for the task of handling a large organization, and dealing with the kind of large-scale operations which went on continuously. The whole system involved about 250 miles (400km) of conduit, most of it underground, but some on *substructiones* (p. 38) and some on arches. Frontinus obviously visited almost every part of this system in person, and gives his equivalent of map-references for the location of the sources (e.g. 'The intake of the Aqua Claudia is on a turn-off to the left from the Sublacensian Way at the 38th milestone, about 300 paces along'). He gives exact measurements of the conduits in 'paces'* even down to half a pace in two cases. Every part of this system had to be inspected at regular intervals, and a considerable work-force was kept permanently employed in maintaining it.

This work-force consisted of two contingents of slaves. One (numbering about 240) was maintained from the public funds, supplemented from the water-rates charged to private consumers. The other, numbering about 460, was the personal property of the emperor.

Frontinus was therefore in charge of a total labour force of about 700, including overseers, 'reservoir-keepers', stonemasons, plasterers and others. At the start of his duties he had not only to see to the renovation of various parts of the system which had fallen into disrepair, but also to get back some members of his work force who, as a result of bribery, had been taken off their proper work and put onto odd jobs for private individuals. What is more, the income from the water rates, which should have been used to support the 'public' force, had been diverted into the private funds of the emperor Domitian.

For all these tasks, it must have been difficult to find anyone more admirably suited than Frontinus. His seniority and authority gave him the power to check corruption and raised him above any need to involve himself in it. In his military commands he was used to dealing with many thousands of troops, and all the problems of supply, finance and administration which that involved. But above all, his keen interest and deep sense of duty, which made him research his job, both in the history books, the legal records (which he quotes extensively) and 'on the ground', gave him an understanding which must have commanded respect in all his subordinates, and fear in those with guilty consciences.

And if he strikes us as perhaps a little ponderous, and a little obsequious towards his emperor, two points should be remembered. The formulae he uses at the start of his work were polite

*The pace *(passus)* was 5 Roman feet, i.e. 4 ft 10.2 in, or 1.4785 m. The Roman mile *(mille passum)* was thus 1616 yds 2 ft, or 0.9185 statute miles (1.4785 km).

conventions, and had no more real meaning than (say) 'I remain, Sir, your obedient servant'. And just occasionally he shows a touch of ironical Roman humour. He tells us (Bk II, Chapter 115) that the official in charge of branch-pipe connections who had allowed illegal pipes to be connected up underground (in return for a bribe) was known as '*a punctis*'. The joke consists in giving a high-sounding official name to an illegal activity, as one might say in English, 'senior commissioner for water-theft in the Ministry of Punctures'.

PLINY

His full name was Gaius Plinius Secundus, and he is known as 'Pliny the Elder', to distinguish him from his nephew Gaius Plinius Caecilius Secundus, whose letters survive in some quantity.

Pliny the Elder belonged to the social class known as 'knights' *(equites)*, which meant that, though educated and well-to-do, his family had not previously aspired to high political office. He was born in A.D. 23, and began his career with a spell in the army, as an officer in the cavalry in Germany from A.D. 47–57. During the remainder of the reign of Nero (i.e. until A.D. 68) he stayed in obscurity, probably owing to his distrust of the Emperor, but under Vespasian (who became Emperor in 69) he was given various military commands and finally made the 'chief of the fleet' *(praefectus classis)* based at Misenum on the Bay of Naples. This was an administrative post, which carried a great deal of responsibility for ship building and maintenance, supplies, finance, etc.

This post was, indirectly and by coincidence, the cause of his death. All his life he had a passionate interest in natural philosophy, especially the more impressive and unusual phenomena of nature. He was with the fleet at Misenum when the great eruption of Vesuvius occurred in August A.D. 79., and he insisted on putting out in a small boat to observe it from closer quarters. At some risk to himself, he put into the shore to pick up and evacuate some survivors, but was himself overcome by the sulphurous fumes and died.

He was apparently a man of immense energy and indefatigable zeal. In a letter to a friend (III, 5) the younger Pliny tells how his uncle used to get up in the morning before dawn, and work incredibly long hours. He had the ability to take brief 'cat-naps' during the working day, and a short siesta after his midday meal (a light snack) enabled him to 'start the day afresh.' He was of course very much occupied with official business, and in his early and middle years with a legal career, but every moment of his leisure was spent in voracious reading of innumerable volumes by many authors, both Greek and Latin. He had the habit (laudable in a professional scholar, but no doubt rather annoying) of always making notes and excerpts from everything he read (or rather, everything as it was read to him by a slave secretary). He used those as the basis for his extensive writings on cavalry warfare, biography, history, oratory, and (this alone of all his works has survived) natural philosophy. The Latin title is *Naturalis Historia,* the word *historia* having its older Greek meaning of 'enquiry' or 'research'.

It is in 37 books (i.e. 37 rolls of papyrus in the original manuscript text) and contains, as he proudly boasts in the preface, 'more than 20,000 important facts, drawn from 100 principal authors'. The first 'book' is really an index of topics and sources; the subjects are then dealt with as follows:

Book 2	The universe (stars, planets, phenomena, astronomy).
Books 3–6	Descriptive geography of the world.
Book 7	Human anatomy and physiology.
Books 8–11	Zoology, descriptive.
Books 12–19	Botany; descriptive accounts of plant structure, seeds, reproduction, etc.
Books 20–27	Medical substances derived from plants, their curative properties.
Books 28–32	Medical substances derived from animals.
Books 33–37	Mineral substances (metals, stones, etc.), mining, metallurgy and the use of derived substances in medicine, painting and architecture.

Pliny used a wide variety of sources in compiling his encyclopaedia, and he was wont to say (again, according to his nephew) that 'no book was so bad that he could not get some benefit from it'. This attitude, charitable and likeable though it may be, had unfortunate consequences. The quality of his work varies enormously, from the first-rate (when he is following good sources and even more so, when he can bring his own experience to bear) to the deplorable, when he is not in a position to see the worthlessness of his source, or has been unable to interpret it correctly. Apart from this variation in quality, Pliny has two characteristics which sometimes strike the reader as regrettable. He has a great fondness for digression and anecdote, which leads him away from his central topic from time to time, and he also moralizes at length on certain topics, particularly wealth and extravagance, which he attacks with a puritanical zeal whenever the opportunity arises.

However, his work was of great value to the Middle Ages and Renaissance as a storehouse of the knowledge of antiquity on a wide range of subjects, and as such, it is of great value also to the historian of science.

Appendix

The Reconstruction of a Trireme

When I wrote in the first edition 'To build replicas of a *pentekonter* and a trireme, and to establish figures by experiment, would be prohibitively expensive', I could hardly have foreseen that within ten years a reconstruction of a trireme was to be designed, built and rowed in sea trials held, most appropriately, in the Aegean sea in the vicinity of Salamis.

The original initiative for this project came from a Cambridge classical scholar, Professor John Morrison, who over many years had gathered together and interpreted all the evidence from ancient sources, and John Coates, a former Chief Naval Architect for the Ministry of Defence, who translated the evidence into working designs from which a replica could be constructed. A Trireme Trust was formed and originally it was hoped that finance could be raised and the vessel built in Britain; but at an early stage the Greek government became interested in the project, both for its own sake as a piece of research into their own naval history, and also as a publicity symbol to assist in their claim to host the Olympic Games. This second aim was not achieved, but they none the less persevered with the project, and the ship was built; it was named *Olympias,* and commissioned in the Greek Navy in 1987. A large sum of Greek taxpayers' money (somewhere in the region of £750,000) was spent on the construction and thereafter very generous provision was made for volunteer men and women rowers from the Trireme Trust to visit and be accommodated in the naval base at Poros, where the ship was based, and take part in sea trials.

In an Appendix on this scale, it is clearly impossible to go into much detail of the ship and its performance; so I have gathered under a few headings what might be called second thoughts on, and amendments to, my text in chapter 6, (pp.133–51 and the Appendix on pp.166–9).

THE DIMENSIONS AND ARRANGEMENT OF THE HULL

The vital evidence on the overall size of the trireme (the authentic Greek spelling is *trieres*) came from the excavation of shipbuilders' sheds (see p. 145); from this it was established that the overall length could not have been more than about 40m (130ft) and its beam at the widest point no more than about 6m (20ft). But these dimensions do not define the cross-section profile of the vessel, nor do those ancient illustrations which show the ship in side elevation. The shape on the right-hand side of my Fig.52 (p.145) needs correction. The outrigger (*parexeiresia*) did not project sharply from the hull, but was supported on curving brackets which formed an extension of the 'wine-glass' curvature of the hull; also, my diagram shows a bottom which is much too rounded, and which ignores the downward projection of the keel.

It might be thought that the task of drawing up a design from the meagre evidence available would be impossibly difficult; but John Coates makes it clear that on many occasions he was faced with a problem which was difficult to solve, but for which in the end there was only one possible solution. For example, it is known from various sources that there were 27 rowers on each side in the lower two banks, and 31 in the top bank. It is also known that the horizontal distance between rowers was 'two cubits' (taken for the *Olympias* design to be 88.8 cm), so the positions of the beams (*zyga*) on which the middle tier rowers (*zygioi*) sat are predetermined within quite narrow limits. There is one important difference between the arrangement of the three tiers of oarsmen suggested on my p. 144 and that adopted for the *Olympias*. I suggested that the middle tier (*zygioi*) rowed with the pegs against which their oars were pulled ('tholepins') fixed on the gunwale. (I use that familiar term loosely — the trireme did *not* carry cannon!) On the *Olympias* the *zygioi* are seated lower down, and they row through oarports though, unlike those of the bottom tier (*thalamioi*) they do not need leather bags (*askomata*) around their oar-shafts to keep the water out when the sea is rough, or when the ship heels over on a sharp turn. Even so, neither of the two lower tiers can actually see the blades of their oars; they have to learn by practice the exact height to which their hands must be raised to dip the blades into the water (the 'catch') and how low they must go for the blade to be lifted clear for the 'recovery'. They also have to learn to gauge from a very limited field of vision exactly what the oarsmen in their immediate vicinity are doing, so as to avoid the risk of the oars clashing together.

Another problem for the rowers is a chronic lack of space for movement. To obtain the best performance each oarsman should be able to reach forward with his (please assume 'or her' throughout) arms straight at the catch, and move from the hips, using additional force from the leg muscles; this is of course greatly helped by the sliding seats of modern racing eights. But on the *Olympias* the seats are fixed, and the total horizontal movement of the rower's hands is limited to about 33½ins (85cm); this means that any rower more than about 5ft 8ins tall (1.72m) cannot straighten his arms at the catch without hitting the back of the rower in front of him. So instead of relaxing his arm muscles, he has to waste energy in keeping them taut; and if his body and leg movements are not exactly co-ordinated, he is left with only his arm muscles to pull the oar-loom hard against his chest at the finish. As if this were not enough, the unfortunate *thalamioi* in the lowest tier have a 3½ins square horizontal beam at about eye-level in front of them and another behind their heads, which they have to crouch to avoid at each stroke. I believe that this goes a long way to explaining the discrepancy between J.E. Gordon's estimate (see p.167) of the maximum possible speed at 11.5 knots, and the 8.2 knots actually achieved by a Trireme Trust crew on one or two occasions. The higher estimate, which is now thought to be unrealistic, was based on an assumed power output of $\frac{1}{3}$ h.p. (248 watts) per man at 70% propulsive efficiency; these are quite reasonable figures in themselves, but do not take into account the limitations of human muscle-contraction, which in turn limit the maximum speed of movement of the oar-blades. A crew of ancient Greek rowers, none of whom was over 5ft 6ins in height, and who had trained together as a team for months or even years, might well have achieved 9.5 or even 10 knots, whereas the volunteer crews, many of them well over 6ft tall, were unable to use their full potential. During one speed trial a reduced crew of 121 (in effect, the upper two tiers of rowers) managed to develop just under $\frac{1}{5}$ h.p. per rower, and to reach a speed of 8.2 knots for a short period. Considering that most of them had no experience of rowing anything remotely like a trireme, and only two to three weeks to learn that very difficult art, their performance was most creditable.

Incidentally, should anybody reproach the Greeks for not having invented sliding seats, a moment's thought would show that they would not have improved the trireme's performance much. The horizontal distance between each rower and the next would have to be increased by at least 25%; in John Coates' view, a carvel-constructed hull like that of the *Olympias* could not be made any longer without increasing its beam and draught; that would increase the water resistance and affect its manoeuvrability, thus (probably) offsetting all the advantages of the extra power available.

THE CONSTRUCTION OF THE HULL

'Carvel' construction is described on pp. 136–7. When the decision was taken to build the ship in Greece, the contract was given to a firm of shipbuilders who, inevitably, had no experience whatever of this technique. As the types of wood used in ancient triremes are nowadays difficult to obtain in the Eastern Mediterranean, the nearest equivalents available from elsewhere were used — Iroko instead of oak for the principal structural members of the hull, the keel, and the timbers of the stem and ram, and oak only for the tenons joining the planks and for the dowels (*gomphoi*) which held them in place . Douglas fir (*Pseudotsuga Menziesii*) was used for the shell of the hull instead of a Mediterranean pine or fir, and also for the oars used in the first trials; this raised a problem, as it is not a true fir, but a rather heavier wood, which made rowing more difficult. Even so, the hull of *Olympias* has 'positive buoyancy', and would not sink if holed (see p. 148).

It would, of course, have been a counsel of perfection (or perhaps lunacy) to insist that the shipbuilders should use the ancient tools of their craft, so electric power tools were freely used at all stages. Cutting each mortise in the facing edges of the planks was done in antiquity by drilling round holes in a row and chiselling them out to the rectangular shape; since they are spaced out at intervals of about 92 mm (3½ins), there are about 20,000 of them in the hull, and the task clearly required some automation. But even modern technology did not solve one particular problem — that of 'hogging'.

This is a phenomenon which arises whenever a ship's hull has 'sharp ends'. The water in which it is immersed exerts an upward thrust on its bottom which is greatest amidships and considerably less at the bows and stern; the ship can be pictured as a beam which is supported in the middle and which, unless it is stiff enough, sags at each end. If the tenons (or 'wedges', see pp. 136–7) fit very tightly, the hull will be quite stiff; but if there is any slackness in the mortises the planks can slip sideways, allowing the hull as a whole to bend upwards or downwards. During the construction of *Olympias* it was noticed that some of the tenons did not fit as tightly as they should, and a resin glue was injected into the gaps to remedy this. Despite this treatment, after the ship was launched the hull actually hogged by about 55mm. ($2\frac{1}{4}$ins) — that is, the bow and stern dropped by that distance below their original positions relative to the centre. This still remains a problem, made worse by the fact that the hogging is not equal on the port and starboard sides.

THE PERFORMANCE OF THE *OLYMPIAS*

Of course, the performance figure which grabs the headlines and gets into the record books is the top speed of a ship; but in the real conditions of ancient naval warfare it was the control of the speed and the manoeuvrability which really counted; the need for continuous high speed only arose when the ship was pursuing a retreating enemy or itself trying to escape from an engagement. An interesting analogy (though it has serious limitations) can be drawn between the trireme and the fighter planes of World War II. Both had weaponry which operated only in the forward direction (the fixed cannon of the Spitfire and Hurricane and the ram of the trireme) so that the whole ship or plane itself had to be aimed at the target, and the attack timed very precisely. Also, for the most part the target plane or ship had no effective counter-weaponry, so the only tactic available to it was rapid evasive action. In such a situation, the ability to accelerate suddenly and to make a 'tight turn' without loss of speed are the most important factors; the Spitfire and the trireme scored highly on both points.

The weapon of surprise was very important in ancient sea warfare (see p.151) and a sudden attack on ships which had been beached, and were not yet in formation to fight, could be very effective. The time taken to get the volunteer crew aboard *Olympias* and put the oars out 'without haste' was just over 10 minutes using one gangplank, and could have been halved by using two. A Hellenic Navy crew (who, unlike the volunteers, were subject to chivvying by their officers) managed the same manoeuvre in 1½ minutes, and likewise the reverse one — shipping their oars and coming ashore at the double. Another interesting figure is the time taken to 'abandon ship' (jump overboard) in an emergency. The Greek sailors achieved this in 24 seconds; it could have meant the difference between life and death if enemy marines boarded a damaged vessel.

More important than maximum speed was the ability to make a sharp turn, to accelerate or decelerate and, when the ram penetrated the hull of an enemy ship, to reverse quickly and withdraw it. All these manoeuvres were successfully carried out in the trials; rowed at full speed the trireme could make a 90° turn in 30 seconds, while travelling about twice its own length, and a 'right-about' in one minute. Acceleration, too, was very good; in one of the earliest trials (in 1987) it was taken from 0–3 knots in 10 seconds, to about 4.7 knots in 20 seconds, and to a maximum of about 6.2 knots in 30 seconds. Quite understandably (though much to the disappointment of the volunteer crews) the vessel was not allowed to engage in any ramming exercises; but a number of reversing manoeuvres were carried out. After some unsuccessful attempts in the early trials, it was found possible to back water, and accelerate the ship from zero to about three knots astern in quite a short time. Of course, it was not enough merely to extract the ram from the target; the ship had to be re-positioned for the next attack, and may have gone 'full speed astern' for some distance.

The *Olympias* is at present (August 1999) displayed on dry land in a section of the National Maritime Museum at Palaion Faliron (Piraeus); she has contributed very greatly to our understanding of the trireme, for which we must be grateful to the Greek government and people.

Some Further Thoughts

CHAPTER 1

p. 13.   Since I wrote (in 1977) of the 'difficulties of gearing' in cranes which use internal combustion engines, these have been entirely solved by hydraulic transmission.

p. 25.   The problem of the Moselle stone-saws has been ingeniously solved by D.L. Simms (*Technology and Culture* 24, 1983). The 'screeching saws' which were 'dragged through' the stones were endless ropes, pulled round by water-wheels and sprinkled with abrasive sand as they passed through the stone.

CHAPTER 2

Some notes on lead pipes (pp. 42–8).

(a) The dangers of lead poisoning from water-pipes were not as great as in modern systems. If the water is at all chalky a deposit (calcium carbonate) quickly forms on the inside of the pipe which effectively isolates the water from the lead; moreover, the water has to be in stationary contact with the lead for contamination to take place, whereas throughout the entire Roman system the moving water ran past the stationary lead all the time.

(b) I am still sceptical about the strength of lead pipes, and their ability to stand high internal pressures. There is an interesting passage in Horace (*Epistles* I, 10, 20–1) in which he says (ironically) 'I suppose that the water which strives to burst the lead (*tendit rumpere plumbum*) in a small town street is purer than that which tumbles down the course of a mountain stream.' This shows, even in a poetic text, a general awareness both of bursting pipes and of lead pollution.

(c) One other small point — a query this time. Soldering irons must have been essential for pipe assembly; but have any been

found? They should be easily identifiable, with copper or bronze
bits riveted to iron shafts, and traces of tin and lead on their tips.
The wooden handles might not survive.

(d) p. 47.   There are, it seems, clear traces of lead in the soil
along the line of the pipe at Pergamon; for want of any alternative, we have to believe that lead pipes would have been able to
stand a pressure of 260 lb/sq. in. (nowadays referred to as '18.5
bar'). Their total disappearance was no doubt due to recycling.

(e) Taps (p.52).   I was taken to task by Hodge (p. 322) over
these remarks; but I thought it would be clear enough to the reader
that I was referring only to 'discharge' taps such as we have over
baths and sinks (known in the U.S.A., I believe, as faucets). Hodge
describes a number of taps, none of which has the downward-curv-
ing spout which is absolutely essential for a 'discharge tap'. They
are all 'stop-cocks' to be fitted in supply pipes.

CHAPTER 3

The remains of a late 18th- or early 19th-century screw pump are
now on display in the Kennet and Avon Canal Museum at Devizes;
there is a photograph of it in use in the 1920's.

An interesting fact emerged from experiments with a *cochlias*
reconstructed by engineering students at Reading; there is a maximum speed limit, beyond which the centrifugal force keeps the
water stationary in the pump, reducing the load on the power
source and causing it to race away at high speed.

The Silchester pump (p.79), one found at Metz in 1905 and another found near Tarrant Hinton in Dorset in 1985 all suggest that
these pumps were made entirely of wood, apart from the piston-
heads which were clad with iron and incorporated leather washers.

CHAPTER 5

Dr Michael Lewis (see bibliography) has constructed and tested a
scale model of an *onager*, and Dr V.G. Hart has proposed a mathematical model for it, with equations of motion (which are not for
the innumerate or the faint-hearted!). Other interesting experimen-
tation, most of it as yet unpublished, has been carried out. Digby
Stevenson has succeeded in making spring-rope from animal sinew

(see *Journal of the Society of Archer Antiquaries* Vol. 40, 1997) and has made and tested a *cheiroballistra* (see p. 130) with remarkable performance figures: Dr David Sim has made a *gastraphetes* (see pp. 99–104) with a powerful modern bow, and found it remarkably easy to load.

Chapter 6

p. 136.   Drawing on Lionel Casson's work, I used the word 'keelson'. A correspondent (Barry Chapman) tells me that this is the American use of the word; apparently the English term for that element in a ship's hull is 'hog'.

p. 138.   I have revised my opinion on the *hypozomata;* (a) they were *inside* the hull, and (b) they were tightened in rough seas, not to prevent the planks from springing apart but to prevent hogging (see Appendix — The Reconstruction of a Trireme).

Chapter 8

On p.188 I referred to 'our' system of 24 equinoctial hours, implying that it is a modern development; in fact it was used by the Hellenistic astronomers as early as the third century B.C.

Chapter 9

Much more information is now readily available from the sourcebook by Humphrey, Oleson and Sherwood (see bibliography).

Hero of Alexandria: Under pressure from the original Series Editor my criticism of Benjamin Farrington's *Greek Science* (on pp. 202–3) was muted; the book has long been superseded by much better-informed work, but some of the Marxist fog still clings to Hero. A glance at his *Mechanics* shows that he was *not* (as Hodge alleges, p. 201) an impractical academic. I have reconstructed a number of his devices (including his steam engine, the linear-flow siphon and the 'primer') and they all work perfectly well; his machine for cutting a female thread in a wood block has also been reconstructed and tested. It is of course necessary to have, or to be properly provided with, the necessary knowledge of materials and techniques of construction.

Two brief comments on Hero's devices:

(a) The trick jug described on p. 202 has re-appeared recently, in the form of a divided bottle containing an Irish liqueur in one half and cream in the other, which can be poured out in any desired proportions by opening or closing the air inlets.

(b) The 'five-drachma piece' used to work the slot-machine (p. 203) was probably a bronze version of the silver coin, larger and heavier but worth much less.

Bibliography

GENERAL WORKS

Much the most important is J.W. Humphrey, J.P. Oleson and A.N. Sherwood, *Greek and Roman Technology: a Sourcebook* (Routledge, London 1997)

See also:

K.D. White, *Greek and Roman Technology* (Pub. Cornell University Press, London, 1984)

J.G. Landels art. 'Engineering' in *Civilization of the Ancient Mediterranean*, ed. Grant & Kitzinger (Charles Scribner's Sons, New York 1988) Vol. I pp.323–52. (Covers some topics not dealt with in this book)

CHAPTER 1

There is no general work which covers this range of subjects. There are some interesting discussions of man-power in the reports of sea trials of the trireme (see ch. 6)

On animal power see:

Ann Hyland, *Equus—the Horse in Roman Times* (Pub. Routledge, London 1990)

J. Spruytte, *Early Harness Systems* (London 1983)
    on the 'ox-powered' ship:

E.A. Thompson, *A Roman reformer and inventor* (O.U.P., 1952)

On water-wheels see:

Hodge (next chapter) pp.254–61

M.J.T. Lewis, *Millstone and Hammer— the Origins of Water Power* (Hull University Press, 1997)

Örjan Wikander 'Water-mills in Ancient Rome' in *Opuscula Romana* xii:2 (1979) 13–36.

CHAPTER 2

There is now a definitive textbook on this subject — A. Trevor Hodge, *Roman Aqueducts and Water Supply* (Duckworth, London 1992), containing a very detailed review of all the archaeological evidence, with copious interpretation and a very full bibliography.

See also:

N.A.F. Smith, 'Attitudes to Roman engineering and the question of the inverted siphon', in *History of Technology* Vol. 1 (1976) 45–71.

On the organization and administration see:

Janet Delaine, 'The Baths of Caracalla' in *Journal of Roman Archaeology*, Supp.25 (Portsmouth, Rhode Island 1997)

CHAPTER 3

This subject also has a definitive textbook — John Peter Oleson, *Greek and Roman Mechanical Water-Lifting Devices: the History of a Technology* (University of Toronto Press, 1984). It has very detailed accounts of all the devices known at that date and a very full bibliography.

CHAPTER 4

There is no major new work on this subject, but see J.P. Oleson, 'A Roman sheave block from the harbour of Caesarea Maritima, Israel' in *Int. Journal of Nautical Archaeology and Underwater Exploration* 12 (1983) 155–70.

CHAPTER 5

E.W. Marsden's two volumes *Greek and Roman Artillery*, Vol. I *Historical Development* and *Technical Treatises* (O.U.P. 1969 & 1971 respectively) have been reprinted (Sandpiper Reprints, PostScript 1997), and remain the fullest and most detailed account available, though a number of his conclusions have been challenged.

See also V.G. Hart and M.J. Lewis 'The mechanics of the onager', in *Journal of Engineering Mathematics 20:* 345–65 (Dordrecht, 1986)

CHAPTER 6

Lionel Casson, *Ships and seamanship in the ancient world* (Princeton U.P., 1971) remains the best general work on the subject; see also his *Ships and Seamanship in Ancient Times* (College Station, Texas & London 1994)

There is now an excellent definitive textbook on the trireme, J.S. Morrison and J.F. Coates, *The Athenian Trireme* (C.U.P. 1986). It presents a new and very detailed study of the evidence, and of the technical problems involved in reconstructing the vessel. It was published before the completion of the *Olympias,* but a new edition by Boris Rankov, which will include a full account of the sea trials and a very full bibliography, is due to be published early in 2000.

Detailed (and at times technical) reports of the *Olympias* sea trials have been published:

1987 (ed. Morrison & Coates) BAR International Series 486, Oxford 1989

1988 (ed. Coates/Platis/Shaw) Oxbow Monograph no. 2 Oxford 1990

1990 (ed. Shaw) Oxbow Monograph no. 31 Oxford 1993

CHAPTER 7

(See the three entries under 'animal power' in Chapter 1)

Alison Burford, 'Heavy Transport in Antiquity', in *Economic History Review XIII* (1960), 1–8

On the warehouses at Ostia see H.P. Rickman, *Roman Granaries and Store Buildings* (C.U.P. 1971)

For a complete survey of the archaeological evidence for prehistoric wheeled vehicles 'from the Atlantic Coast to the Caspian Sea' see Stuart Piggott, *The Earliest Wheeled Transport* (Cornell University Press, 1983)

CHAPTER 8

There is an extensive and very detailed study of the development of Greek scientific thought in G.E.R. Lloyd, *The Revolutions of Wisdom* (University of California Press, Berkeley 1987)

For a concise but comprehensive account there are two volumes by the same author, *Greek Science, Thales to Aristotle* and *Greek Science after Aristotle* (Chatto & Windus, London 1970 and 1973)

CHAPTER 9

The ancient writers on technological subjects have not attracted as much attention as the technology itself. There is still no English edition of Hero's Greek text (only the German one of a century ago in the Teubner series) and there is an urgent need for a scholarly edition of Vitruvius. It would require a scholar (or a team) with expert knowledge of Latin and Greek and skill in palaeography: the oldest manuscripts abound in corruptions, having been copied by scribes who had no understanding of the content, and Granger, in his Loeb edition (Heinemann, London 1931, 1934) followed a method of textual criticism which was very severely castigated, not without good cause, by Housman. It would also require an expert in the theory of Greek architecture, on which Vitruvius draws extensively, and someone with the necessary mechanical knowledge to deal with the machines. A mathematician/astronomer would be needed to unravel the mysteries of the *analemma*, and a skilled illustrator could contribute much to the reader's understanding. Some of these skills may be in very short supply in the next century.

Index